Molecular Kinetics in Condensed Phases

Molecular Kinetics in Condensed Phases

Theory, Simulation, and Analysis

Ron Elber
W.A. "Tex" Moncrief Jr. Endowed Chair
Oden Institute for Computational Engineering and Sciences
and Department of Chemistry
The University of Texas at Austin
USA

Dmitrii E. Makarov
The University of Texas at Austin
USA

Henri Orland
Institut de Physique Théorique
CEA Saclay
France

This edition first published 2020
© 2020 John Wiley & Sons Ltd

The right of Ron Elber, Dmitrii E. Makarov, and Henri Orland to be identified as the authors of this work has been asserted in accordance with law.

Registered Offices
John Wiley & Sons, Inc., 111 River Street, Hoboken, NJ 07030, USA
John Wiley & Sons Ltd, The Atrium, Southern Gate, Chichester, West Sussex, PO19 8SQ, UK

Editorial Office
The Atrium, Southern Gate, Chichester, West Sussex, PO19 8SQ, UK

For details of our global editorial offices, customer services, and more information about Wiley products visit us at www.wiley.com.

Wiley also publishes its books in a variety of electronic formats and by print-on-demand. Some content that appears in standard print versions of this book may not be available in other formats.

Library of Congress Cataloging-in-Publication Data

Names: Elber, Ron, author. | Makarov, Dmitrii E., author. | Orland, Henri, author.
Title: Molecular kinetics in condensed phases : theory, simulation, and analysis / Ron Elber, Dmitrii E. Makarov, Henri Orland.
Description: First edition. | Hoboken, NJ : Wiley, [2020] | Includes bibliographical references and index.
Identifiers: LCCN 2019024971 (print) | LCCN 2019024972 (ebook) | ISBN 9781119176770 (hardback) | ISBN 9781119176787 (adobe pdf) | ISBN 9781119176794 (epub)
Subjects: LCSH: Chemical kinetics–Mathematical models. | Stochastic processes–Mathematical models. | Molecular structure. | Molecular theory.
Classification: LCC QD502 .E43 2019 (print) | LCC QD502 (ebook) | DDC 541/.394–dc23
LC record available at https://lccn.loc.gov/2019024971
LC ebook record available at https://lccn.loc.gov/2019024972

Cover Design: Wiley
Cover Images: Courtesy of Ron Elber; © atakan/iStock.com

Set in 10/12pt WarnockPro by SPi Global, Chennai, India

Printed and bound in Singapore by Markono Print Media Pte Ltd

10 9 8 7 6 5 4 3 2 1

We dedicate this monograph to our families, in appreciation of their love and support.
Henri Orland dedicates the book to Elisabeth, Chloé, Jonathan, Sarah, and Yasha.
Dmitrii E. Makarov to Valentina, Evgenii, and Vsevolod Makarov.
Ron Elber to Virginia, Dassi, Nurit ,and Nir.

Contents

Acknowledgments

RE acknowledges research support from the Robert A. Welch Foundation (Grant No. F-1896) and from the National Institutes of Health (Grant No. GM059796). He is also grateful to his co-workers Alfredo E Cardenas, Arman Fathizadeh, Piao Ma, Katelyn Poole, Clark Templeton, and Wei Wei, who have provided many constructive comments during the preparation of this text. The assistance of Arman Fatizadeh and Atis Murat in the preparation of the cover image and of Alfredo E. Cardenas and Wei Wei in production of some of the figures is greatly acknowledged. The Mueller trajectories were computed using a program written by Piao Mao and the alanine dipeptide calculations were run by Wei Wei.

DEM acknowledges research support from the Robert A. Welch Foundation (Grant No. F-1514) and the National Science Foundation (Grant No. CHE 1566001), as well as the illuminating comments from his colleagues Alexander M. Berezhkovskii, Peter Hamm, Erik Holmstrom, Hannes Jonsson, Eduardo Medina, Daniel Nettels, Eli Pollak, Rohit Satija, Benjamin Schuler, and Flurin Sturzenegger.

HO would like to thank M. Bauer and K. Mallick for numerous illuminating discussions.

Introduction: Historical Background and Recent Developments that Motivate this Book

This book grew from the lectures given at the summer schools that the three of us organized in Telluride and in Lausanne. The purpose of the schools was to introduce young researchers to modern kinetics, with an emphasis to applications in life sciences, in which rigorous methods rooted in statistical physics are playing an increasing role. Indeed, molecular-level understanding of virtually any process that occurs in cellular environment requires a kinetic description as well as the ability to measure and/or compute the associated timescales. Similarly, dynamic phenomena occurring in materials, such as nucleation or fracture growth, require a kinetic framework. Experimental and theoretical developments of the past two decades, which are briefly outlined below, rejuvenated the well-established field of kinetics and placed it at the juncture of biophysics, chemistry, molecular biology, and materials science; the composition of the class attending the schools, which included graduate students and postdocs working in diverse areas, reflected this renewed interest in kinetic phenomena. In this book we strive to introduce a diverse audience to the modern toolkit of chemical kinetics. This toolkit enables prediction of kinetic phenomena through computer simulations, as well as interpretation of experimental kinetic data; it involves methods that range from atomistic simulations to physical theories of stochastic phenomena to data analysis.

The material presented in this book can be loosely divided into three interconnected parts. The first part consists of Chapter 1–7 and provides a comprehensive account of stochastic dynamics based on the model where the dynamics of the relevant degree(s) of freedom is described by the Langevin equation. The Langevin model is especially important for several reasons: First, it is the simplest model that captures the essential features of any proper kinetic theory, thereby offering important insights. Second, it often offers a minimal, low-dimensional description of experimental data. Third, many problems formulated within this model can be solved analytically. Theoretical approaches to Langevin dynamics described in this book include the Fokker–Planck equation, mapping between stochastic dynamics and quantum mechanics, and Path Integrals.

The second part of the book, comprised of Chapter 8–12, describes rate theories and explains our current understanding of what a "reaction rate" is and how it is connected to the underlying microscopic dynamics. The rate theories discussed in this part range from the simple transition state theory to multidimensional theories of diffusive barrier crossing. While they were originally formulated in different languages and in application to different phenomena, we explain their interrelationship and formulate them using a unified language.

Finally, the third part, formed by Chapter 13–19, focuses on atomistic simulations and details approaches that aim to: (i) perform such simulations in the first place, (ii) predict kinetic phenomena occurring at long time timescales that cannot be reached via brute force simulations, and (iii) obtain insights from the simulation data (e.g. by inferring the complex dynamical networks or transition pathways in large-scale biomolecular rearrangements).

We start with a brief historical overview of the field. To most chemists, the centerpiece of chemical kinetics is the Arrhenius law. This phenomenological rule states that the rate coefficient of a chemical reaction, which quantifies how fast some chemical species called reactant(s) interconvert into different species, the product(s), can be written in the form

$$k = v e^{-\frac{E_a}{k_B T}} \tag{1}$$

To someone with a traditional physics background, however, this rule may require much explaining. In classical mechanics, the state of a molecular system consisting of N atoms is described by its position in phase space $(\mathbf{x}, \mathbf{p})$, where $\mathbf{x}$ is the $3N$-dimensional vector comprised of the coordinates of its atoms and $\mathbf{p}$ the corresponding vector of their momenta. The time evolution in phase space is governed by Newton's second law, which results in a system of $6N$ differential equations. What exactly does a chemist mean by "reactants" or "products"? When, why, and how can the complex Newtonian dynamics (or the dynamics governed by the laws of quantum mechanics) be reduced to a simple rate law, what is the physical meaning of v and E_a, and how can we predict their values?

Early systematic attempts to answer these questions started in the early twentieth century with work by Rene Marcelin [1] and culminated with the ideas of transition state theory formulated by Wigner [2], Eyring [3] and others (see, e.g., the review [4]. In the modern language (closest to Wigner's formulation) "transition state" is a hypersurface that divides the phase space into the reactants and products; it further possesses the hypothetic property that any trajectory crossing this hypersurface heading from the reactants to the products will not turn back. Interestingly, the existence (in principle) of such a hypersurface has remained an unsettled issue even in the recent past [5]. In practice, transition state theory is often a good approximation for gas-phase reactions involving few atoms, or for transitions in solids, which have high symmetry. But practical transition-state theory estimates of the prefactor v for biochemical phenomena that take place in solution are usually off by many orders of magnitude.

Several important developments advanced our understanding of the Arrhenius law later in the twentieth century. First, Kramers in his seminal paper [6] approached the problem from a different starting point: instead of considering the dynamics in the high-dimensional space involving all the degrees of freedom, he proposed a model where one important degree of freedom characterizing the reaction is treated explicitly, while the effect of the remaining degrees of freedom is captured phenomenologically using theory of Brownian motion. Kramers's solution of this problem described in Chapter 6 of this book has had a tremendous impact, particularly, on biophysics.

Second, Keck and collaborators [7] showed that, even if the hypersurface separating the products from the reactants does not provide a point of no return and can be recrossed by molecular trajectories, it is still possible to correct for such recrossings, and – if there are not too many – practical calculation methods exist that will yield the exact rate [8–10].

Third, Kramers's ideas were extended beyond simple, one-dimensional Brownian dynamics to include multidimensional effects [11] and conformational memory [12, 13]. Moreover, a unification of various rate theories was achieved [4, 14, 15] using the idea that Brownian dynamics can be obtained from the conservative dynamics of an extended system where the degree of freedom of interest is coupled bilinearly to a continuum of harmonic oscillators [16, 17]. This unified perspective on rate theories will be explored in detail in Chapter 9–11 of this book.

There is, however, more to kinetics than calculating the rate of an elementary chemical step. Many biophysical transport phenomena cannot be characterized by a single rate coefficient as in Eq. (1). Yet a description of such phenomena in full atomistic detail would both lack insight (i.e. fail to provide their salient features or mechanisms) and be prohibitive computationally. The challenge is to find a middle ground between simple phenomenological theories and expensive molecular simulations, a task that has been tackled by a host of new methods that have emerged in the last decade and that employ "celling" strategies described in Chapters 18–19 of this book.

Another recent development in molecular kinetics has been driven by single-molecule experiments, whose time resolution has been steadily improving: it has become possible to observe properties of molecular transition paths by catching molecules *en route* from reactants to products [18, 19] as they cross activation barriers. Such measurements provide critical tests of various rate theories described in Chapters 6, 10, 11, and inform us about elusive reaction mechanisms. Properties of transition paths, such as their temporal duration and dominant shape are discussed in Chapter 6–7.

Given that the focus of this book is on the dynamic phenomena in condensed phases, and especially on biophysical applications, a number of topics were left out. Those, for example, include theories of reaction rate in the gas phase. Likewise, the low-friction regime (also known as the energy diffusion regime) and the related Kramers turnover problem [15] for barrier crossing are not covered here, as those are rarely pertinent for chemical dynamics in solution. Throughout most of this book, it was assumed that the dynamics of molecules can be described by the laws of Newtonian mechanics, with quantum mechanics only entering implicitly through the governing inter- and intramolecular interactions. This assumption holds in many cases, but the laws of quantum mechanics become important when a chemical reaction involves the transfer of a light particle, such as proton or electron, and/or when it occurs at a very low temperature. The subject of quantum rate theory could fill a separate book [20, 21]; here its discussion is limited to a single chapter (Chapter 12) that is only meant to provide a rudimentary introduction.

References

1 Marcelin, R. (1915). *Ann. Phys.* 3: 120.

2 Wigner, E. (1932). *Z. Phys. Chem. Abt. B* 19: 203.

3 Eyring, H. (1935). The activated complex in chemical reactions. *J. Chem. Phys.* 3: 107.

4 Pollak, E. and Talkner, P. (2005). Reaction rate theory: what it was, where is it today, and where is it going? *Chaos* 15 (2): 26116.

5 Mullen, R.G., Shea, J.E., and Peters, B. (2014). Communication: an existence test for dividing surfaces without recrossing. *J. Chem. Phys.* 140 (4): 041104.

6 Kramers, H.A. (1940). Brownian motion in a field of force and the diffusion model of chemical reactions. *Physica* 7: 284–304.

7 Shui, V.H., Appleton, J.P., and Keck, J.C. (1972). Monte Carlo trajectory calculations of the dissociation of HCl in Ar. *J. Chem. Phys.* 56: 4266.

8 Chandler, D. (1978). *J. Chem. Phys.* 68: 2959.

9 Bennett, C.H. (1977). Molecular dynamics and transition state theory: the simulation of infrequent events. In: *Algorithms for Chemical Computations*, vol. 46, 63–97. American Chemical Society.

10 Truhlar, D.G., Garrett, B.C., and Klippenstein, S.J. (1996). Current status of transition-state theory. *J. Phys. Chem.* 100 (31): 12771–12800.

11 Langer, J.S. (1969). *Ann. Phys.* (N.Y.) 54: 258.

12 Grote, R.F. and Hynes, J.T. (1980). The stable states picture of chemical reactions. II. Rate constants for condensed and gas phase reaction models. *J. Chem. Phys.* 73 (6): 2715–2732.

13 Grote, R.F. and Hynes, J.T. (1981). Reactive modes in condensed phase reactions. *J. Chem. Phys.* 74 (8): 4465–4475.

14 Pollak, E. (1986). Theory of activated rate-processes - a new derivation of Kramers expression. *J. Chem. Phys.* 85 (2): 865–867.

15 Hanggi, P., Talkner, P., and Borkovec, M. (1990). 50 years after Kramers. *Rev. Mod. Phys.* 62: 251.

16 Zwanzig, R. (2001). *Nonequilibrium Statistical Mechanics*. Oxford University Press.

17 Caldeira, A.O. and Leggett, A.J. (1983). Quantum tunneling in a dissipative system. *Ann. Phys.* 149: 374.

18 Chung, H.S. and Eaton, W.A. (2018). Protein folding transition path times from single molecule FRET. *Curr. Opin. Struct. Biol.* 48: 30–39.

19 Neupane, K., Foster, D.A., Dee, D.R. et al. (2016). Direct observation of transition paths during the folding of proteins and nucleic acids. *Science* 352 (6282): 239–242.

20 Benderskii, V.A., Makarov, D.E., and Wight, C.A. (1994). *Chemical Dynamics at Low Temperatures*. New York: Wiley.

21 Nitzan, A. (2006). Chemical Dynamics in Condensed Phases: Relaxation, Transfer and Reactions in Condensed Molecular Systems. Oxford/New York: Oxford University Press; p xxii, 719 p.

1

The Langevin Equation and Stochastic Processes

1.1 General Framework

In this section, we will discuss how to describe a system of interacting particles in contact with a heat bath (or thermal reservoir) at temperature T. Examples of such systems are countless. Just to cite a few, we will often refer to a protein in a solvent, the binding of a protein to a DNA, etc. Our approach closely follows the original work of Paul Langevin [1]. Very complete presentations of the subjects treated in this part can be found in the following books [2]. For a thorough review of the mathematics of stochastic processes, we refer the reader to the book by Gardiner [3].

Consider a single particle of mass m in a fluid. In the following, we will assume the physical space to be one-dimensional, since generalization to arbitrary dimension is straightforward. The formalism developed here may apply to some *reaction coordinates* of the system. Reaction coordinates are low-dimensional collective variables that represent the evolution of a system along a transformation pathway (physical or chemical).

Going back to the one-dimensional case, in the fluid, the particle is subject to a friction force F_f given by *Stokes law* [4]

$$F_f = -\gamma v \qquad (1.1)$$

where v is the velocity of the particle and γ is the friction coefficient. The *Newton equation* for the particle is thus

$$m\frac{dv}{dt} = F_f = -\gamma v \qquad (1.2)$$

which can be integrated as

$$v(t) = v(0)e^{-\frac{\gamma}{m}t} \qquad (1.3)$$

where $v(0)$ is the initial velocity of the particle. In the limit of large time $t \to \infty$, $v(t) \to 0$. Due to the friction force, the particle goes to rest at large time, with relaxation time $\tau_r = m/\gamma$

Molecular Kinetics in Condensed Phases: Theory, Simulation, and Analysis,
First Edition. Ron Elber, Dmitrii E. Makarov and Henri Orland.
© 2020 John Wiley & Sons Ltd. Published 2020 by John Wiley & Sons Ltd.

However, at large time, the particle should reach thermal equilibrium with the heat bath at temperature T, which implies that the probability distribution for its velocity should be the *Maxwell distribution*

$$P(v) = \left(\frac{m}{2\pi k_B T}\right)^{1/2} e^{-\beta \frac{mv^2}{2}} \tag{1.4}$$

where $\beta = 1/k_B T$ and k_B is the Boltzmann constant. A consequence of the Maxwell distribution is the *equipartition theorem*, which states that at thermal equilibrium

$$\frac{1}{2}m\langle v^2 \rangle = \frac{1}{2}k_B T \tag{1.5}$$

This equipartition theorem is just a consequence of the Gaussian integral

$$\langle v^2 \rangle = \int_{-\infty}^{+\infty} dv \left(\frac{m}{2\pi k_B T}\right)^{1/2} e^{-\beta \frac{mv^2}{2}} v^2 = \frac{k_B T}{m} \tag{1.6}$$

This is obviously not compatible with Eq. (1.3) which implies a vanishing velocity at long time. To overcome this difficulty, Langevin proposed to add a time-dependent random force $\xi(t)$ to those acting on the particle, to account for the thermal agitation (Brownian motion) and for the collisions with the heat bath molecules at temperature T. The Newton equation thus becomes

$$m\frac{dv}{dt} = -\gamma v + \xi(t) \tag{1.7}$$

The random force $\xi(t)$ is also called *random noise*. The average of this force should vanish, as it represents random thermal agitation with no specific direction or intensity

$$\langle \xi(t) \rangle = 0 \tag{1.8}$$

for any t. The bracket means an average over all possible noise histories, that is over a large number of evolutions of the same particle, with different realizations of the noise.

Taking the average of Eq. (1.7) over different random noises, we have

$$m\frac{d\langle v \rangle}{dt} = -\gamma \langle v \rangle \tag{1.9}$$

which is the same as Eq. (1.2) above and implies $\langle v(+\infty) \rangle = 0$ at large time. The integration of Eq. (1.7) is straightforward and yields

$$v(t) = v_0 e^{-\frac{\gamma}{m}t} + \frac{1}{m}\int_0^t d\tau e^{-\frac{\gamma}{m}(t-\tau)}\xi(\tau) \tag{1.10}$$

which at large time $t \to \infty$ becomes

$$v(t) = \frac{1}{m}\int_0^t d\tau e^{-\frac{\gamma}{m}(t-\tau)}\xi(\tau) \tag{1.11}$$

Physically, the random force $\xi(t)$ is due to collisions of the heat bath molecules with the studied particle. Therefore the intensity and direction of the random force changes at each collision, and thus the correlation time τ_c of the random force is of the order of magnitude of the typical timespan between collisions of the heat bath molecules with the particle. If the relaxation time τ_r of the particle and the time scales t at which we study the particle is much larger than the correlation time τ_c, we may neglect the time correlations of the random force and write

$$\langle \xi(t)\xi(t') \rangle = 2\lambda\delta(t - t') \tag{1.12}$$

where $\delta(t - t')$ denotes the Dirac delta function (or distribution), which is non zero only if $t = t'$. We now show how to adjust the coefficient λ in order to satisfy the equipartition theorem at thermal equilibrium (1.5). The variance of the velocity is given by

$$\begin{aligned}
\langle v^2(t) \rangle &= \frac{1}{m^2} \int_0^t d\tau \int_0^t d\tau' e^{-\frac{\gamma}{m}(2t-\tau-\tau')} \langle \xi(\tau)\xi(\tau') \rangle \\
&= \frac{2\lambda}{m^2} \int_0^t d\tau \int_0^t d\tau' e^{-\frac{\gamma}{m}(2t-\tau-\tau')} \delta(\tau - \tau') \\
&= \frac{2\lambda}{m^2} \int_0^t d\tau e^{-\frac{2\gamma}{m}(t-\tau)} \\
&= \frac{\lambda}{m\gamma} \left(1 - e^{-\frac{2\gamma}{m}t} \right)
\end{aligned} \tag{1.13}$$

At large time, we have $\langle v^2(t) \rangle = \frac{\lambda}{m\gamma}$, which from (1.5) implies

$$\frac{1}{2}m\langle v^2 \rangle = \frac{\lambda}{2\gamma} = \frac{1}{2}k_B T \tag{1.14}$$

We thus see that the coefficient λ is related to the friction coefficient γ through the relation

$$\lambda = \gamma k_B T \tag{1.15}$$

We now turn to the general case of a particle in a heat bath at temperature T on which a force $F(x, t)$ (possibly time-dependent) is acting. Again, adding the Stokes force, the Newton equation reads

$$m\frac{dv}{dt} + \gamma v = F \tag{1.16}$$

If the force is conservative, it is the gradient of a potential energy $F = -\frac{dU}{dx}$ and we have

$$m\frac{dv}{dt} + \gamma v + \frac{dU}{dx} = 0 \tag{1.17}$$

Multiplying this equation by the velocity $v = \frac{dx}{dt}$

$$mv\frac{dv}{dt} + \gamma v^2 + v\frac{dU}{dx} = \frac{d}{dt}\left(\frac{1}{2}mv^2 + U\right) + \gamma v^2$$
$$= \frac{dE}{dt} + \gamma v^2$$
$$= 0 \tag{1.18}$$

where $E = \frac{1}{2}mv^2 + U$ is the total energy of the system. In the above equation, we have used the fact that for a time independent potential, $\frac{dU(x(t))}{dt} = \frac{dU(x(t))}{dx}\frac{dx}{dt} = v\frac{dU(x(t))}{dx}$. Equation (1.18) implies $\frac{dE}{dt} = -\gamma v^2 \leq 0$ and therefore the energy of the system decreases with time. When $t \to +\infty$, if the energy is bound below, the system will fall into a local minimum of the potential energy $U(x)$, with a vanishing velocity. In this state, $\frac{dU}{dx} = 0$ and $v = 0$ and Eq. (1.17) is satisfied.

As before, the equipartition theorem is not satisfied, and to overcome this problem, we again introduce a random force $\xi(t)$ and obtain the *Langevin equation*

$$m\frac{d^2x}{dt^2} + \gamma\frac{dx}{dt} = F + \xi(t) \tag{1.19}$$

where

$$\langle \xi(t) \rangle = 0$$
$$\langle \xi(t)\xi(t') \rangle = 2\gamma k_B T \delta(t - t') \tag{1.20}$$

These equations do not specify higher moments of the random force. In the following, we will assume that the force is *Gaussian distributed*, with correlation functions given by (1.20). A particle subject to such a random force is called a *Brownian particle*. In this equation, the term $\gamma\frac{dx}{dt}$ term is the friction term, the force term F is called the *drift term* and the noise term $\xi(t)$ is called the *diffusion term*.

Equation (1.19) is called a *stochastic differential equation* (SDE) and constitutes a very active branch of modern mathematics. This equation describes a *Markov process*, that is a stochastic process such that the state of the system at time $t + dt$ depends (stochastically) only on the state of the system at time t.

A random noise such as in Eq. (1.20) with constant variance and zero correlation time is called a *white noise*. A noise with non-zero correlation time is called *colored noise* and we will see an example of that in section 1.2 on the Ornstein-Uhlenbeck process. To be complete, let us mention that we have discussed only the case when the friction term is instantaneous. In some systems, the friction force does not take the instantaneous Stokes form, but may depend on the previous history of the system. In section 1.8, we will briefly describe the case when the friction force is expressed in terms of a memory kernel. We will also discuss the case when the particle moves in a inhomogeneous solvent, where the diffusion coefficient D depends on x.

1.2 The Ornstein-Uhlenbeck (OU) Process

Following the previous section, we consider a particle in the absence of an external force F. Eq. (1.19) becomes

$$m\frac{d^2x}{dt^2} + \gamma\frac{dx}{dt} = \xi(t) \tag{1.21}$$

where the noise is Gaussian, with moments given by Eq. (1.20). The above equation can be expressed in a simpler form by writing it in terms of the velocity v

$$m\frac{dv}{dt} + \gamma v = \xi(t) \tag{1.22}$$

The stochastic motion of the particle is called an *Ornstein-Uhlenbeck process* (OU). The solution of Eq. (1.22) is given by

$$v(t) = v_0 e^{-\frac{\gamma}{m}t} + \frac{1}{m}\int_0^t d\tau\, e^{-\frac{\gamma}{m}(t-\tau)}\xi(\tau) \tag{1.23}$$

where v_0 is the initial velocity of the particle and $\xi(t)$ is a white Gaussian noise with correlations given by Eq. (1.20). By integrating this equation we get

$$x(t) = x_0 + \frac{v_0 m}{\gamma}\left(1 - e^{-\frac{\gamma}{m}t}\right) + \frac{1}{m}\int_0^t d\tau \int_0^\tau d\tau'\, e^{-\frac{\gamma}{m}(\tau-\tau')}\xi(\tau') \tag{1.24}$$

where x_0 is the initial position of the particle. Let us calculate the correlation time of the velocity of the particle (in absence of any external force). The average velocity is given by

$$\langle v(t)\rangle = v_0 e^{-\frac{\gamma}{m}t} + \frac{1}{m}\int_0^t d\tau\, e^{-\frac{\gamma}{m}(t-\tau)}\langle\xi(\tau)\rangle$$

$$= v_0 e^{-\frac{\gamma}{m}t}$$

since the average noise vanishes (from Eq. (1.20)). The velocity correlation function is given by

$$\langle(v(t) - \langle v(t)\rangle)(v(t') - \langle v(t')\rangle)\rangle = \langle v(t)v(t')\rangle_c$$

$$= \frac{1}{m^2}\int_0^t d\tau \int_0^{t'} d\tau'\, e^{-\frac{\gamma}{m}(t+t'-\tau-\tau')}\langle\xi(\tau)\xi(\tau')\rangle \tag{1.25}$$

$$= \frac{2\gamma k_B T}{m^2}\int_0^t d\tau \int_0^{t'} d\tau'\, e^{-\frac{\gamma}{m}(t+t'-\tau-\tau')}\delta(\tau - \tau') \tag{1.26}$$

where we use the notation $\langle\ldots\rangle_c$ to denote the *connected* correlation function of the variable, also called the *second cumulant*, where the average of the variable has been subtracted. In the above equation, the δ−function is non-zero only when $\tau = \tau'$, and this can be satisfied if and only if both τ and τ' are smaller than the smallest of t and t'. Let us denote

$$t_< = \inf(t, t')$$
$$t_> = \sup(t, t')$$

where $\inf(t, t')$ and $\sup(t, t')$ denote the smaller (resp. larger) of the two times t and t'. Then $t + t' = t_< + t_>$ and we can write

$$
\begin{aligned}
\langle v(t)v(t')\rangle_c &= \frac{2\gamma k_B T}{m^2} \int_0^{t_<} d\tau\, e^{-\frac{\gamma}{m}(t_< + t_> - 2\tau)} \\
&= \frac{k_B T}{m} \left(e^{-\frac{\gamma}{m}(t_> - t_<)} - e^{-\frac{\gamma}{m}(t_> + t_<)} \right) \\
&= \frac{k_B T}{m} \left(e^{-\frac{\gamma}{m}|t - t'|} - e^{-\frac{\gamma}{m}(t + t')} \right)
\end{aligned}
\tag{1.27}
$$

At large times, $t, t' \to \infty$, the correlation function decays exponentially with $|t - t'|$:

$$\langle v(t)v(t')\rangle_c \sim \frac{k_B T}{m} e^{-\frac{\gamma}{m}|t - t'|} \tag{1.28}$$

We thus see that the velocity of an OU process is also a Gaussian noise, but since it has a finite correlation time, it is an example of a coloured Gaussian noise. By setting $t = t'$ in Eq. (1.27), we obtain the variance of the velocity as

$$\langle v^2(t)\rangle_c = \frac{k_B T}{m} \left(1 - e^{-\frac{2\gamma}{m}t} \right) \tag{1.29}$$

At large time, we recover the equipartition theorem. We note that the variance of the velocity is finite, and has a finite limit when $t \to \infty$.

Due to its inertial mass m, a Brownian particle has a tendency to continue in the direction of its velocity, despite the action of the random force. As a result, the velocity of the particle has a finite correlation time, that is, a time over which its direction and length persist. From Eq. (1.28), the correlation time of the velocity is

$$\tau_v = \frac{m}{\gamma} \tag{1.30}$$

which is the time-scale over which the velocity persists. Beyond that time, the velocities of the particles can be considered as independent random variables. As we will show next, this implies that if we are interested in the behavior of the system at time scales larger than m/γ, we can neglect the mass term in the Langevin equation.

It is interesting to compute the probability distribution of the velocity and of the position of the particle. There is a powerful theorem in mathematics which states that if a random variable is a linear combination of some Gaussian random variables, then it is itself a Gaussian distributed random variable, and its distribution is thus entirely specified by its average and its variance (see Appendix A). From Eq. (1.23), we see that the velocity is a linear combination of the Gaussian noise $\xi(t)$ and thus it is Gaussian distributed with distribution

$$P(v,t) = \frac{1}{\mathcal{N}} e^{-\frac{\left(v-v_0 e^{-\frac{\gamma}{m}t}\right)^2}{2\frac{k_BT}{m}\left(1-e^{-\frac{2\gamma}{m}t}\right)}} \rightarrow_{t\to\infty} \frac{1}{\mathcal{N}} e^{-\frac{mv^2}{2k_BT}} \tag{1.31}$$

where $\mathcal{N}$ is the normalization constant. It follows that the velocity of the particle is always bound, and its distribution converges to the Maxwell distribution.

Similarly, the average and variance of the position can be computed. The average is given by

$$\langle x(t)\rangle = x_0 - \frac{v_0 m}{\gamma}\left(e^{-\frac{\gamma}{m}t}-1\right) + \frac{1}{m}\int_0^t d\tau \int_0^\tau d\tau' e^{-\frac{\gamma}{m}(\tau-\tau')}\langle\xi(\tau')\rangle$$

$$= x_0 - \frac{v_0 m}{\gamma}\left(e^{-\frac{\gamma}{m}t}-1\right) \tag{1.32}$$

and the variance is

$$\left\langle(x(t)-\langle x(t)\rangle)^2\right\rangle$$

$$= \frac{1}{m^2}\int_0^t d\tau_1 \int_0^{\tau_1} d\tau_2 \int_0^t d\tau_3 \int_0^{\tau_3} d\tau_4 e^{-\frac{\gamma}{m}(\tau_1-\tau_2)}e^{-\frac{\gamma}{m}(\tau_3-\tau_4)}\langle\xi(\tau_2)\xi(\tau_4)\rangle$$

$$= \frac{2\gamma k_BT}{m^2}\int_0^t d\tau_1 \int_0^{\tau_1} d\tau_2 \int_0^t d\tau_3 \int_0^{\tau_3} d\tau_4 e^{-\frac{\gamma}{m}(\tau_1-\tau_2)}e^{-\frac{\gamma}{m}(\tau_3-\tau_4)}\delta(\tau_2-\tau_4) \tag{1.33}$$

Again, since τ_2 must be equal to τ_4, they are both limited by the smallest of the two times τ_1 and τ_3. We denote

$$\tau_< = \inf(\tau_1,\tau_3)$$
$$\tau_> = \sup(\tau_1,\tau_3)$$

With this notation, Eq. (1.33) becomes

$$\left\langle(x(t)-\langle x(t)\rangle)^2\right\rangle = \frac{2\gamma k_BT}{m^2}\int_0^t d\tau_1 \int_0^t d\tau_3 \int_0^{\tau_<} d\tau_2 e^{-\frac{\gamma}{m}(\tau_1+\tau_3-2\tau_2)}$$

$$= \frac{k_BT}{m}\int_0^t d\tau_1 \int_0^t d\tau_3 \left(e^{-\frac{\gamma}{m}|\tau_1-\tau_3|}-e^{-\frac{\gamma}{m}(\tau_1+\tau_3)}\right)$$

$$= 2Dt - 2\frac{Dm}{\gamma}\left(1-e^{-\frac{\gamma}{m}t}\right) \tag{1.34}$$

where $D = \frac{k_BT}{\gamma}$ is a constant. At large time, we recover the standard diffusion law

$$(\Delta x)^2 \sim 2Dt \tag{1.35}$$

where $(\Delta x)^2 = \langle (x(t) - \langle x(t) \rangle)^2 \rangle$. Eq. (1.35) implies that D is the diffusion coefficient of the particle and we have the *Einstein relation*

$$D = \frac{k_B T}{\gamma} \tag{1.36}$$

which relates the diffusion and the friction coefficients [5]. At short times $t \ll \frac{m}{\gamma}$, we may expand Eq. (1.34) to second order in t and we obtain

$$(\Delta x)^2 \sim \frac{D\gamma}{m} t^2 \tag{1.37}$$

which shows that the movement of the particle is ballistic at short times.

Knowing the average and variance of the position, we can calculate the probability distribution of the position. For large time, we obtain

$$P(x,t) = \frac{1}{\mathcal{N}} \exp\left(-\frac{\left(x - x_0 - \frac{v_0 m}{\gamma}\right)^2}{4Dt}\right) \tag{1.38}$$

Due to the inertia of the particle, to its initial velocity v_0 and to the friction force, the particle remains centered around $x_0 + \frac{v_0 \gamma}{m}$ and diffuses around that position.

1.3 The Overdamped Limit

From the preceding section, it appears that if we are interested in time scales larger than the correlation time $\tau_v = m/\gamma$, the velocities can be considered as random independent variables. As we will now show, this amounts to neglect the mass term in the Langevin equation. Indeed, taking the Fourier transform of the Langevin Equation (1.19), we have

$$(-m\omega^2 + i\gamma\omega)\tilde{x}(\omega) + \tilde{W}(\omega) = \tilde{\xi}(\omega) \tag{1.39}$$

where

$$\tilde{x}(\omega) = \int_{-\infty}^{+\infty} dt \, e^{-i\omega t} x(t)$$

$$\tilde{\xi}(\omega) = \int_{-\infty}^{+\infty} dt \, e^{-i\omega t} \xi(t)$$

$$\tilde{W}(\omega) = \int_{-\infty}^{+\infty} dt \, e^{-i\omega t} \frac{\partial U(x(t))}{\partial x} \tag{1.40}$$

If we are interested in the long time limit of the system, only the small ω Fourier components become relevant. For small enough ω, the mass term carries an ω^2

term and can be neglected compared to the friction term which is linear in ω. The Langevin equation thus reduces to

$$\gamma \frac{dx}{dt} = F + \xi(t) \tag{1.41}$$

with the noise satisfying equations Eq. (1.20). This equation is called the *overdamped Langevin equation*, or *Brownian dynamics* (BD). Dividing this equation by γ and using Eq. (1.36), this equation is often written as

$$\frac{dx}{dt} = D\beta F + \eta(t) \tag{1.42}$$

$$= \frac{1}{\gamma} F + \eta(t) \tag{1.43}$$

with

$$\langle \eta(t) \rangle = 0$$
$$\langle \eta(t)\eta(t') \rangle = 2D\delta(t - t') \tag{1.44}$$

where

$$D = \frac{k_B T}{\gamma} \tag{1.45}$$

The timescale beyond which the mass term is irrelevant can be read off Eq. (1.39). Indeed, for ω sufficiently small, we have

$$m\omega^2 \ll \gamma\omega \tag{1.46}$$

which in real time translates into

$$\tau \gg \frac{m}{\gamma} \tag{1.47}$$

Therefore, for time scales larger than $\tau_v = \frac{m}{\gamma}$, which turns out to be the correlation time of velocities from Eq. (1.30), one can neglect the mass term in the Langevin Equation (1.19) and replace it by the overdamped Langevin Equation (1.41). Let us show that in the overdamped limit, the diffusion relation Eq. (1.35) is satisfied. Indeed, in the case of no potential $U = 0$, we can integrate Eq. (1.41) and obtain

$$x(t) = \frac{1}{\gamma} \int_0^t d\tau \xi(\tau) \tag{1.48}$$

where we have assumed that the particle is at point 0 at time 0. Therefore, using Eq. (1.20), the correlation function of the position is given by

$$
\begin{aligned}
\langle x(t)x(t')\rangle &= \frac{1}{\gamma^2} \int_0^t d\tau \int_0^{t'} d\tau' \langle \xi(\tau)\xi(\tau')\rangle \\
&= \frac{2k_B T}{\gamma} \int_0^t d\tau \int_0^{t'} d\tau' \delta(\tau - \tau') \\
&= \frac{2k_B T}{\gamma} \inf(t, t')
\end{aligned}
\tag{1.49}
$$

and the square displacement is given by

$$
\langle x^2(t)\rangle = \frac{2k_B T}{\gamma} t \tag{1.50}
$$

We thus recover the Einstein relation (1.36). It follows that the short time scale τ_v beyond which the overdamped limit is valid is given by

$$
\tau_v = \frac{m}{\gamma} = \frac{MD}{RT} \tag{1.51}
$$

where M is the molar mass of the solute, and $R \simeq 8.31 \text{J/mol/K}$ is the ideal gas constant. For example, for an amino-acid in water with average molar mass $M = 110 \text{g/mol}$, and diffusion coefficient at room temperature typically $10^{-5} \text{cm}^2/\text{s}$ we obtain $\tau_s = 5.10^{-14}$ s. Although typically in molecular dynamics simulations, the discretization timestep is of the order of femtoseconds (10^{-15} s), in many instances one is not interested in details of the dynamics at timescales below 10^{-13} s, and thus Brownian dynamics will be appropriate.

To be complete, let us note that for a system with N degrees of freedom (e.g. a many-particle system in 3d), the Langevin Equation takes the form

$$
\gamma_i \frac{dx_i}{dt} = F_i + \xi_i(t) \tag{1.52}
$$

with

$$
\begin{aligned}
&\langle \xi_i(t)\rangle = 0 \\
&\langle \xi_i(t)\xi_j(t')\rangle = 2\gamma_i \delta_{ij} k_B T \delta(t - t')
\end{aligned}
\tag{1.53}
$$

where the diffusion and friction coefficients D_i, γ_i with

$$
D_i = \frac{k_B T}{\gamma_i}
$$

may depend on the nature of the degree of freedom i.

1.4 The Overdamped Harmonic Oscillator: An Ornstein–Uhlenbeck process

When studying the motion of a system near a local minimum, it is reasonable to expand the potential to second order around it and approximate it by a harmonic potential. Consider a particle in a harmonic potential

$$U(x) = \frac{1}{2} K x^2 \tag{1.54}$$

with spring constant K. The overdamped Langevin equation is

$$\frac{dx}{dt} = -D\beta K x + \eta(t) \tag{1.55}$$

We see that this equation is formally equivalent to Eq. (1.22), and thus the overdamped motion of a particle in a harmonic potential is an OU process. It follows that all the results of section 1.2 can be adapted. Defining $\omega = D\beta K$, we obtain

$$\langle x(t) \rangle = x_0 e^{-\omega t}$$
$$\langle x(t) x(t') \rangle_c = \frac{k_B T}{K} (e^{-\omega |t-t'|} - e^{-\omega(t+t')})$$
$$\langle x^2(t) \rangle_c = \frac{k_B T}{K} (1 - e^{-2\omega t}) \tag{1.56}$$

which shows that the motion of the particle remains bound, with a correlation time equal to

$$\tau_c = \frac{1}{\omega} = \frac{k_B T}{KD} \tag{1.57}$$

From Eq. (1.56), we see that the relaxation time τ_r of the system to equilibrium is also given by

$$\tau_r = \tau_c = \frac{1}{\omega} \tag{1.58}$$

The p.d.f of the particle is Gaussian at all time, given by

$$P(x, t) = \frac{1}{\mathcal{N}} \exp\left(-\frac{\beta K (x - x_0 e^{-\omega t})^2}{2(1 - e^{-2\omega t})} \right) \tag{1.59}$$

and converges to the Boltzmann distribution when $t \to \infty$. In Figure 1.1 we show the plot of a typical OU trajectory in the potential $U = x^2/2$ at temperature $T = 0.1$ with timestep $dt = 0.001$ and total duration $t = 50$. As can be seen, the motion of x is confined around $x = 0$.

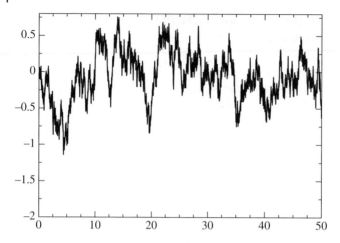

Figure 1.1 Ornstein-Uhlenbeck trajectory.

1.5 Differential Form and Discretization

In most of the following, we will stick to the overdamped Langevin equation, or Brownian Dynamics (BD). In differential form, the overdamped Langevin equation (1.41) can be written as

$$dx = \frac{F}{\gamma}dt + dB(t) \tag{1.60}$$

where $dB(t) = \eta(t)dt$ is the differential of a Brownian process $B(t)$. A Brownian process is a continuous stochastic process, whose variations during any infinitesimal time interval dt are independent normal Gaussian variables of zero mean and variance proportional to dt. According to Eq. (1.44), we have

$$dB(t) = B(t + dt) - B(t) = \sqrt{2Ddt}\,\mathcal{N}(0,1) \tag{1.61}$$

where $\mathcal{N}(0,1)$ is a Gaussian normal variable of 0 mean and unit variance. We can write symbolically

$$P(dB(t)) = \frac{1}{\sqrt{4\pi Ddt}}e^{-\frac{(dB(t))^2}{4Ddt}} \tag{1.62}$$

This equation describes an *Itô diffusion*, and has been thoroughly studied in mathematics. Equations (1.60) and (1.61) show the fundamental result that the typical displacement of the particle during time dt is of order $\sqrt{dt}$

$$dx \sim \sqrt{dt} \tag{1.63}$$

The velocity of the particle $\frac{dx}{dt} \sim \frac{1}{\sqrt{dt}}$ is discontinuous at all time. In fact, since the drift term $\frac{F}{\gamma}dt$ is of order dt, taking the square of Eq. (1.60) and keeping terms to order dt yields

$$(dx)^2 = (dB(t))^2 \tag{1.64}$$

Similarly, the full Langevin Eq. (1.19) can be written in differential form using the momentum $p = m\frac{dx}{dt}$

$$dx = \frac{p}{m}dt$$

$$dp + \frac{\gamma}{m}pdt = Fdt + dB_1(t) \tag{1.65}$$

where

$$dB_1(t) = \sqrt{2\gamma k_B T dt}\,\mathcal{N}(0,1) \tag{1.66}$$

In that case, Eq. (1.63) is replaced by

$$dp \sim \sqrt{dt} \tag{1.67}$$

and the acceleration of the particle is discontinuous at all time.

There is an infinite number of ways to discretize Eq. (1.60) or (1.65), at order dt. A theorem on stochastic processes states that the continuous limit of the Itô diffusion exists, and is independent of the discretization of the equation. We define the timestep dt and the sequence of times

$$\tau_n = ndt \tag{1.68}$$

with

$$t = Ndt \tag{1.69}$$

Integrating Eq. (1.60) between τ_n and $\tau_{n+1} = \tau_n + dt$, we have

$$x_{n+1} = x_n + \frac{1}{\gamma}\int_{\tau_n}^{\tau_{n+1}} F(x(s), s)ds + \int_{\tau_n}^{\tau_{n+1}} dB(s) \tag{1.70}$$

where $x_n = x(\tau_n)$ and $x_{n+1} = x(\tau_{n+1})$. The stochastic term in (1.70) is a sum (integral) of independent Gaussian random variables $dB(s)$. It is thus a Gaussian random variable with

$$\eta_n = \int_{\tau_n}^{\tau_{n+1}} dB(s) \tag{1.71}$$

$$= \int_{\tau_n}^{\tau_{n+1}} ds\eta(s) \tag{1.72}$$

We have

$$\langle \eta_n \rangle = 0 \tag{1.73}$$

$$\langle \eta_m \eta_n \rangle = \int_{\tau_m}^{\tau_{m+1}} ds \int_{\tau_n}^{\tau_{n+1}} ds' \langle \eta(s)\eta(s') \rangle \tag{1.74}$$

$$= \frac{2k_B T}{\gamma} \int_{\tau_m}^{\tau_{m+1}} ds \int_{\tau_n}^{\tau_{n+1}} ds' \delta(s - s') \tag{1.75}$$

If $m \neq n$, the two integration intervals are disjoint and the δ−function vanishes. Therefore

$$\langle \eta_m \eta_n \rangle = 2D\delta_{mn} dt \tag{1.76}$$

This last equation shows that the noise η_n is of order $\sqrt{dt}$. To make this scaling explicit, we introduce the variables $\zeta_n = \frac{\eta_n}{\sqrt{2Ddt}}$ which are normal Gaussian random variables with 0 mean, unit variance

$$\langle \zeta_n \rangle = 0 \tag{1.77}$$

$$\langle \zeta_n^2 \rangle = 1 \tag{1.78}$$

and probability distribution

$$P(\zeta_n) = \frac{1}{\sqrt{2\pi}} e^{-\frac{\zeta_n^2}{2}} \tag{1.79}$$

The integral of the force $F(x(s), s)$ over the interval $[\tau_n, \tau_{n+1}]$ is of order $\tau_{n+1} - \tau_n = dt$. We may thus write it as

$$\int_{\tau_n}^{\tau_{n+1}} F(x(s), s) ds = \Phi_n dt \tag{1.80}$$

where Φ_n depends on the trajectory $x(s)$ between times τ_n and τ_{n+1}.

Equation (1.70) takes the form

$$x_{n+1} = x_n + \frac{1}{\gamma} \Phi_n dt + \sqrt{2Ddt}\, \zeta_n \tag{1.81}$$

The various discretization schemes of the Langevin equation correspond to different expressions for Φ_n, which are discussed below.

A crucial point to note in Eq. (1.81) is that the force term Φ_n, also called the drift term, is of order dt since it is multiplied by dt, while the random force term, also called the diffusion term, is of order $\sqrt{dt}$. For small timestep, the diffusion term is thus always much larger than the drift term and to dominant order in dt, we have

$$\langle (x_{n+1} - x_n)^2 \rangle \sim 2Ddt \tag{1.82}$$

It is also clear from this equation that the motion of the particle is a Markov process, since its position x_{n+1} at time $\tau_{n+1} = \tau_n + dt$ depends stochastically (through ζ_n) on the state x_n at time τ_n only, and not on the previous ones. Another crucial remark is that x_n (at time τ_n) depends on all ζ_m with

$m = 0, 1, ..., n - 1$ but not on the noise $\zeta_{m'}$ for $m' = n, n + 1, ...,$ that is, not on the random force at times larger or equal to τ_n. This *causality principle* implies that for any function f, we have

$$\langle f(x_n)\zeta_n \rangle = \langle f(x_n)\rangle\langle \zeta_n \rangle = 0 \tag{1.83}$$

In the continuous limit, this is equivalent to

$$\langle f(x(t))dB(t)\rangle = \langle f(x(t))\rangle\langle dB(t)\rangle = 0 \tag{1.84}$$

To conclude this section, we mention the important result that

$$(dB(t))^2 = 2Ddt \tag{1.85}$$

This result is to be understood in terms of stochastic integrals. Indeed, as we show in Appendix B, for any function $f(x)$, we have

$$\int_0^t (dB(s))^2 f(x(s)) = 2D \int_0^t ds f(x(s)) \tag{1.86}$$

Taking a derivative of the above equation with respect to t, we have

$$(dB(t))^2 f(x(t)) = 2Ddt f(x(t)) \tag{1.87}$$

which implies the symbolic identity (1.85).

1.5.1 Euler-Maruyama Discretization (EMD) and Itô Processes

The most common discretization scheme of Eq. (1.41, 1.60) is the Euler-Maruyama method. It amounts to approximate the integral of Eq. (1.80) by

$$\int_{\tau_n}^{\tau_{n+1}} F(x(s), s)ds \sim F(x_n, \tau_n)dt \tag{1.88}$$

and

$$\int_{\tau_n}^{\tau_{n+1}} F(x(s), s)dx(s) \sim F(x_n, \tau_n)(x_{n+1} - x_n) \tag{1.89}$$

for dt small. The Euler-Maruyama discretization of (1.60) uses the values of the functions at the earlier time τ_n. Since in differential calculus, all quantities are computed to order dt, the time at which the function is evaluated in Eq. (1.88) is irrelevant, as any other time in the interval $[\tau_n, \tau_{n+1}]$ would just add corrections of higher order. With this scheme, Eq. (1.81) can be written as

$$x_{n+1} = x_n + \frac{1}{\gamma}F(x_n, \tau_n)dt + \sqrt{2Ddt}\,\zeta_n \tag{1.90}$$

A great advantage of this discretization is that it provides an **explicit** expression of x_{n+1} as a function of x_n.

The numerical solution of the discretized Langevin Equation (1.90) is straightforward: given an initial starting point x_0 and a timestep dt, generate a Gaussian number with a normal distribution $\mathcal{N}(0,1)$ (see Eq. (1.79)), and compute x_1 using Eq. (1.90). Then, iterate the procedure to obtain a sample trajectory of the particle. As we will see in chapter 2.5, one must generate many trajectories, with different "noise histories", to compute dynamical or thermodynamical averages. Figure 1.2 displays a sample overdamped Langevin trajectory of a particle in a quartic well potential $U(x) = \frac{1}{2}(x^2 - 1)^2$ (Figure 1.3)

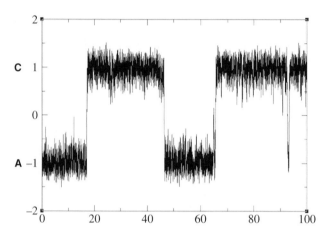

Figure 1.2 Langevin trajectory.

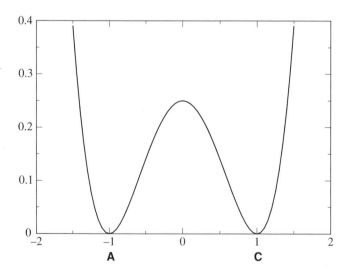

Figure 1.3 Quartic potential.

at temperature $T = 0.1$ with timestep $dt = 0.001$. We clearly see the small thermal fluctuations of the particle in each well, separated by large jumps from one well to the other.

The main drawbacks of the EMD scheme are a strong tendency to instability and slow convergence as $dt \to 0$. Indeed, this scheme is exact to 1st order only. There are many explicit discretization schemes that can improve the accuracy but we won't discuss them here. All these schemes converge to the same continuous limit, called an Itô process. In this process, due to the fact that $dx \sim \sqrt{dt}$, the differential of a function $f(x, t)$ must be expanded to second order as

$$df(x(t), t) = \frac{\partial f}{\partial t} dt + \frac{\partial f}{\partial x} dx + \frac{1}{2} \frac{\partial^2 f}{\partial x^2} (dx(t))^2 \tag{1.91}$$

Using the property Eq. (1.85), we have

$$(dx(t))^2 = (dB(t))^2 = 2D dt \tag{1.92}$$

from which we obtain the *Itô derivative* formula

$$df(x(t), t) = \frac{\partial f}{\partial t} dt + \frac{\partial f}{\partial x} dx + D \frac{\partial^2 f}{\partial x^2} dt \tag{1.93}$$

This formula is of paramount importance in the study of stochastic processes. As a consequence, if the function f does not depend explicitly on time, we have

$$\int_0^T dt \frac{df}{dx} \frac{dx}{dt} = f(x(T)) - f(x(0)) - D \int_0^T dt \frac{d^2 f}{dx^2} \tag{1.94}$$

which, due to the second derivative term, is *not* the standard chain rule of integral calculus.

To be complete, let us mention that the Euler-Maruyama discretization of the full Langevin Equation (1.65) with mass term, is

$$x_{n+1} = x_n + \frac{p_n}{m} dt$$

$$p_{n+1} = p_n - \frac{\gamma}{m} p_n dt + F_n dt + \sqrt{2\gamma k_B T \, dt} \, \zeta_n \tag{1.95}$$

Figure 1.4 displays a sample underdamped Langevin trajectory of a particle of mass $m = 1$ in a quartic well potential $U(x) = \frac{1}{2}(x^2 - 1)^2$ (Figure 1.3) at temperature $T = 0.1$ with a timestep $dt = 0.001$.

Obviously, this trajectory is much smoother than Figure 1.2, due to the presence of the mass term, which regularizes the velocity and makes it continuous. By decreasing the mass, one can show that the trajectory gets rougher and converges to the overdamped case.

1.5.2 Stratonovich Discretization (SD)

A simple way to improve the stability of the discretized equation is to use **implicit** discretizations. These are discretizations where the variable x_{n+1}

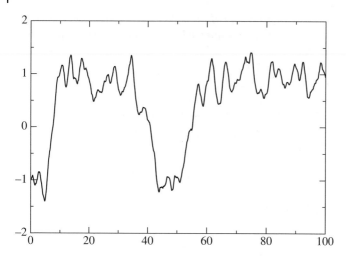

Figure 1.4 Langevin trajectory of a massive particle in a quartic double-well.

depends implicitly on itself. In other words, contrary to an Itô process which depends strictly on the value of the variable in the past, in a Stratonovich process, the variable also depends on its present value.

In the Stratonovich discretization, the integral of Eq. (1.88) is defined as

$$\int_{\tau_n}^{\tau_{n+1}} F(x(s), s) \bullet ds \sim \frac{1}{2}(F(x_{n+1}, \tau_{n+1}) + F(x_n, \tau_n))dt \tag{1.96}$$

and

$$\int_{\tau_n}^{\tau_{n+1}} F(x(s), s) \bullet dx(s) \sim \frac{1}{2}(F(x_{n+1}, \tau_{n+1}) + F(x_n, \tau_n))(x_{n+1} - x_n) \tag{1.97}$$

We use the symbol $\bullet$ to distinguish the Stratonovich integral from the Itô integral. As mentioned in the previous subsection, to order dt, the integral Eq. (1.96) is identical to Eq. (1.88). However, this is not true for Eq. (1.97), since this integral and Eq. (1.89) are both of order $\sqrt{dt}$ and so is their difference. The Stratonovich form implies the following discretization for the Langevin Equation

$$x_{n+1} = x_n + \frac{1}{2\gamma}(F(x_{n+1}, \tau_{n+1}) + F(x_n, \tau_n))dt + \sqrt{2Ddt}\,\zeta_n \tag{1.98}$$

where the random variable ζ_n has the same distribution Eq. (1.79) as in the EMD scheme.

Obviously, the two schemes have the same continuous limit $dt \to 0$. Indeed, the term $F(x_{n+1}, \tau_{n+1})dt$ can be replaced by $F(x_n, \tau_n)dt$, the difference being of higher order in dt (more precisely of order $dt^{3/2}$). Then Eq. (1.98) becomes Eq. (1.90). In fact, all discretizations of the continuous stochastic equation (1.60) which are exact to order dt can be shown to be equivalent when $dt \to 0$. We will see in the next section that this is not the case when the diffusion coefficient D is space dependent.

To conclude this section, we now show that contrary to the Itô scheme, in the Stratonovich scheme, the standard rules of calculus apply. To illustrate this fundamental property, consider the integral of Eq. (1.94), where $f(x)$ is a function of x only. We divide the interval $[0, T]$ in small segments $[\tau_n, \tau_{n+1}]$, with $n = 0, ..., N - 1$. According to (1.97), we have

$$\int_{\tau_n}^{\tau_{n+1}} \frac{df}{dx} \bullet \frac{dx}{dt} dt = \int_{\tau_n}^{\tau_{n+1}} f'(x) \bullet dx(t)$$

$$= \frac{1}{2}(f'(x_{n+1}) + f'(x_n))(x_{n+1} - x_n) \tag{1.99}$$

On the other hand, we have, up to order dt :

$$df(t) = f(x_{n+1}) - f(x_n)$$

$$f(x_{n+1}) = f(x_n) + (x_{n+1} - x_n)f'(x_n) + \frac{1}{2}(x_{n+1} - x_n)^2 f''(x_n) \tag{1.100}$$

$$f(x_n) = f(x_{n+1}) + (x_n - x_{n+1})f'(x_{n+1}) + \frac{1}{2}(x_n - x_{n+1})^2 f''(x_{n+1})$$

$$= f(x_{n+1}) + (x_n - x_{n+1})f'(x_{n+1}) + \frac{1}{2}(x_n - x_{n+1})^2 f''(x_n) \text{ to order } dt \tag{1.101}$$

so that by subtracting Eqs. (1.100) and (1.101), the second order terms cancel and we obtain

$$df = \frac{1}{2}(x_{n+1} - x_n)(f'(x_{n+1}) + f'(x_n)) \tag{1.102}$$

It follows that within the Stratonovich scheme, we have

$$\int_0^T \frac{df}{dx} \bullet \frac{dx}{dt} dt = f(x(T)) - f(x(0)) \tag{1.103}$$

which is the standard chain rule of calculus.

1.6 Relation Between Itô and Stratonovich Integrals

We have seen that the Itô and Stratonovich integral do not satisfy the same rules: the Stratonovich integral satisfies all rules of integral calculus (chain rule, integration by part, etc.) whereas Itô does not. Let us show the relation between the two integrals.

Consider the Itô integral

$$
\begin{aligned}
I_I &= \int_0^{t_f} d\tau \frac{dx(\tau)}{d\tau} f(x(\tau), \tau) \\
&= \int_{x_i}^{x_f} dx(\tau) f(x(\tau), \tau)
\end{aligned}
\tag{1.104}
$$

By definition, this integral is the limit when $N \to \infty$ of

$$
I_I = \sum_{n=0}^{N-1} (x_{n+1} - x_n) f(x_n, \tau_n)
\tag{1.105}
$$

with $x_0 = x_i$ and $x_N = x_f$. Similarly, consider the Stratonovich integral

$$
\begin{aligned}
I_S &= \int_0^{t_f} f(x(\tau), \tau) \bullet \frac{dx(\tau)}{d\tau} d\tau \\
&= \int_{x_i}^{x_f} f(x(\tau), \tau) \bullet dx(\tau)
\end{aligned}
\tag{1.106}
$$

This integral is the limit when $N \to \infty$ of

$$
I_S = \frac{1}{2} \sum_{n=0}^{N-1} (x_{n+1} - x_n)(f(x_{n+1}, \tau_{n+1}) + f(x_n, \tau_n))
\tag{1.107}
$$

Substracting Eq. (1.105) from Eq. (1.107), we obtain

$$
I_S - I_I = \frac{1}{2} \sum_{n=0}^{N-1} (x_{n+1} - x_n)(f(x_{n+1}, \tau_{n+1}) - f(x_n, \tau_n))
\tag{1.108}
$$

and expanding to order dt, we have

$$
\begin{aligned}
f(x_{n+1}, \tau_{n+1}) - f(x_n, \tau_n) = dt \frac{\partial f(x_n, \tau_n)}{\partial \tau_n} &+ (x_{n+1} - x_n) \frac{\partial f(x_n, \tau_n)}{\partial x_n} \\
&+ \frac{1}{2}(x_{n+1} - x_n)^2 \frac{\partial^2 f(x_n, \tau_n)}{\partial x_n^2}
\end{aligned}
$$

The first term and last terms in the r.h.s above are of order dt and when multiplied by $(x_{n+1} - x_n)$ which is of order $\sqrt{dt}$ in Eq. (1.108), they do not contribute. Only the second term proportional to $(x_{n+1} - x_n)$ does contribute, and this implies

$$
I_S - I_I = \frac{1}{2} \sum_{n=0}^{N-1} (x_{n+1} - x_n)^2 \frac{\partial f(x_n, \tau_n)}{\partial x_n}
\tag{1.109}
$$

which using Eq. (1.85) and taking the continuous limit yields the fundamental relation between the two integrals

$$
I_S - I_I = D \int_0^{t_f} d\tau \frac{\partial f(x(\tau), \tau)}{\partial x(\tau)}
\tag{1.110}
$$

1.7   Space Varying Diffusion Constant

In some physical situations, such as the diffusion of a tracer in a inhomogeneous medium, or when looking at a specific monomer in a polymer chain, the diffusion coefficient may depend on the spatial position of the particle. The overdamped Langevin Equation (1.60) takes the form

$$dx = \beta D(x(t))F(x(t))dt + dB(t) \tag{1.111}$$

where

$$dB(t) = \sqrt{2D(x(t))dt}\,\mathcal{N}\,(0, 1) \tag{1.112}$$

and $\mathcal{N}\,(0, 1)$ is a normal Gaussian variable of 0 mean and variance 1. Let us show now that the Euler-Maruyama and Stratonovich discretization give different equations.

In the Euler-Maruyama scheme, we have

$$\int_{\tau_n}^{\tau_{n+1}} D(x(s))F(x(s), s)ds \sim D(x_n)F(x_n, \tau_n)dt \tag{1.113}$$

$$\int_{\tau_n}^{\tau_{n+1}} dB(s) \sim \sqrt{2D(x_n)dt}\,\zeta_n \tag{1.114}$$

so that the discretized form becomes

$$x_{n+1} = x_n + \beta D(x_n)F(x_n, \tau_n)dt + \sqrt{2D(x_n)dt}\,\zeta_n \tag{1.115}$$

whereas in the Stratonovich scheme, we have

$$\int_{\tau_n}^{\tau_{n+1}} D(x(s))F(x(s), s) \bullet ds = \frac{1}{2}(D(x_{n+1})F(x_{n+1}, \tau_{n+1}) + D(x_n)F(x_n, \tau_n))dt$$

$$\sim D(x_n)F(x_n, \tau_n)dt \tag{1.116}$$

since Eqs. (1.113) and (1.116) are equivalent to order dt. On the other hand, from Eq. (1.96), we have

$$\int_{\tau_n}^{\tau_{n+1}} \bullet dB(s) \sim \frac{1}{2}\left(\sqrt{2D(x_n)dt} + \sqrt{2D(x_{n+1})dt}\right)\zeta_n \tag{1.117}$$

and the Stratonovich discretization is thus

$$x_{n+1} = x_n + \frac{\beta}{2}(D(x_{n+1})F(x_{n+1}, \tau_{n+1}) + D(x_n)F(x_n, \tau_n))dt$$

$$+ \frac{1}{2}\left(\sqrt{2D(x_n)dt} + \sqrt{2D(x_{n+1})dt}\right)\zeta_n \tag{1.118}$$

Assuming that the diffusion coefficient is regular enough (differentiable) we have

$$D(x_{n+1}) = D(x_n) + (x_{n+1} - x_n)\frac{\partial D}{\partial x_n} + \cdots \tag{1.119}$$

From Eq. (1.63), we know that

$$x_{n+1} - x_n \sim \sqrt{dt} \qquad (1.120)$$

so that we may expand to first order in $x_{n+1} - x_n$

$$\sqrt{2D(x_{n+1})dt} = \sqrt{2\left(D(x_n) + (x_{n+1} - x_n)\frac{\partial D}{\partial x_n}\right)dt}$$

$$= \sqrt{2D(x_n)dt}\sqrt{1 + (x_{n+1} - x_n)\frac{\partial \log D(x_n)}{\partial x_n}}$$

$$= \sqrt{2D(x_n)dt}\left(1 + \frac{1}{2}(x_{n+1} - x_n)\frac{\partial \log D(x_n)}{\partial x_n}\right)$$

We have to order dt

$$\sqrt{2D(x_{n+1})dt}\,\zeta_n = \sqrt{2D(x_n)dt}\,\zeta_n + \frac{1}{2}\sqrt{2D(x_n)dt}\,\zeta_n(x_{n+1} - x_n)\frac{\partial \log D(x_n)}{\partial x_n}$$
$$(1.121)$$

To order $\sqrt{dt}$, we have

$$x_{n+1} - x_n = \sqrt{2D(x_n)dt}\,\zeta_n \qquad (1.122)$$

and using a rescaled form of Eq. (1.85), we have

$$\sqrt{2D(x_n)dt}\,\zeta_n(x_{n+1} - x_n) = (x_{n+1} - x_n)^2 \qquad (1.123)$$
$$= (dB(t))^2 \qquad (1.124)$$
$$= 2D(x_n)dt \qquad (1.125)$$

Therefore, expanding Eq. (1.118) to order dt, we have

$$x_{n+1} = x_n + \beta D(x_n)\left(\frac{1}{2}(F(x_{n+1}, \tau_{n+1}) + F(x_n, \tau_n)) + \frac{\partial \log D(x_n)}{\partial x_n}\right)dt$$
$$+ \sqrt{2D(x_n)dt}\,\zeta_n \qquad (1.126)$$

$$= x_n + \beta D(x_n)\left(F(x_n, \tau_n) + \frac{\partial \log D(x_n)}{\partial x_n}\right)dt + \sqrt{2D(x_n)dt}\,\zeta_n \qquad (1.127)$$

Eq. (1.127) is equivalent to Eq. (1.126) at order dt. However, we see that in the case of a space dependent diffusion constant, the Stratonovich discretization introduces an additional drift force to the Itô discretization (1.115), proportional to $\frac{\partial \log D(x_n)}{\partial x_n}$. In addition, the noise becomes *multiplicative*, since it enters in the Langevin Equation as the product of a function $D(x(t))$ by the white noise $\zeta(t)$. Due to this additional drift term, the continuous limits of the Itô and of the Stratonovich discretization are not the same. The question which arises then, is: which discretization should one use to model a physical stochastic process?

1.8 Itô vs Stratonovich

Mathematicians favor the Euler-Maruyama-Itô discretization, as it has some nice mathematical properties, in particular, that of being explicit, i.e. allowing a simple computation of the position at time $t + dt$ knowing that at time t. However, in any physical process, the random Gaussian force has a small but finite correlation time. Therefore, the Gaussian force has some regularity over short periods of time. A theorem due to Wong and Zakai [7] states in essence that, if the stochastic process Eq. (1.111) is approached as the limit of a process where the correlation time of the noise goes to 0, then the process converges to a limit process which should be discretized by the Stratonovich scheme. Therefore, for physical systems with space dependent diffusion constant, the preferred physical discretization is the Stratonovich one. Let us emphasize again that in the case of a homogeneous diffusion constant, the Itô and Stratonovich discretizations are equivalent, and we will use both of them.

From a purely technical point of view, the Itô scheme has the advantage of being explicit, and the position at "next time" $x(t + dt)$ depends in a simple way on the position at the "present time" $x(t)$. The disadvantages is that the differential of a function is not given by the standard formula, and the rules of calculus have to be modified. On the other hand, the Stratonovich scheme has the benefits to be more physical for systems with (even weakly) correlated noise, and to satisfy all rules of ordinary calculus. Its disadvantage is that it is an implicit scheme, which does not allow for a simple expression of $x(t + dt)$ as a function of $x(t)$.

1.9 Detailed Balance

In the case where the force is conservative, it is the gradient of a potential, and one can show that the system satisfies *microscopic reversibility*: for any process taking the particle from x to $x + dx$ during time dt, there is a reverse process from $x + dx$ to x, and at thermal equilibrium, the ratio of the probabilities to find the system at x and at $x + dx$ is given by the ratio of the corresponding Boltzmann weights. This is also expressed by saying that the system satisfies *detailed balance*. Let us show this explicitly.

Consider the overdamped motion of a particle, within the Itô scheme. The particle at $x(t)$ at time t moves to $x(t + dt) = x(t) + dx$ at time $t + dt$ under the action of the external force F and of the random force $dB(t)$

$$dx = \frac{1}{\gamma} F(x(t))dt + dB(t) \tag{1.128}$$

It follows that the probability for the particle to be at point $x(t) + dx$ at time $t + dt$ given it was at $x(t)$ at time t is equal to

$$P(x + dx, t + dt|x, t) = \text{Prob}\left\{ dx = \frac{1}{\gamma}F(x)dt + dB(t) \right\} \tag{1.129}$$

where $\text{Prob}\{A\}$ is the probability of event A, and $P(x, t|y, t')$ is the conditional probability for the particle to be at point x at time t given that it was at point y at time $t' < t$.

Using Eq. (1.62), we may write

$$P(x + dx, t + dt|x, t) = \frac{1}{\sqrt{4\pi Ddt}} e^{-\frac{\left(dx - \frac{1}{\gamma}F(x)dt\right)^2}{4Ddt}} \tag{1.130}$$

Similarly, the probability for the particle to be at x at time $t + dt$ given that it was at $x(t) + dx$ at time t is

$$P(x, t + dt|x + dx, t) = \text{Prob}\left\{ dx = -\frac{1}{\gamma}F(x + xd)dt - dB(t) \right\} \tag{1.131}$$

which translates into

$$P(x, t + dt|x + dx, t) = \frac{1}{\sqrt{4\pi Ddt}} e^{-\frac{\left(dx + \frac{1}{\gamma}F(x + dx)dt\right)^2}{4Ddt}} \tag{1.132}$$

Taking the ratio of the two conditional probabilities above, we obtain (to order dt)

$$\frac{P(x + dx, t + dt|x, t)}{P(x, t + dt|x + dx, t)} = e^{\frac{1}{\gamma D}F(x)dx} \tag{1.133}$$

and using the relation $\gamma D = k_B T$ and the fact that $F(x)dx = dW$ is the work of the force over the displacement dx we have

$$\frac{P(x + dx, t + dt|x, t)}{P(x, t + dt|x + dx, t)} = e^{\frac{dW}{k_B T}} \tag{1.134}$$

If the force is conservative, we have $F = -\frac{\partial U}{\partial x}$ and $dW = -dU$ so that

$$\frac{P(x + dx, t + dt|x, t)}{P(x, t + dt|x + dx, t)} = e^{-\frac{dU}{k_B T}} \tag{1.135}$$

which proves that the Langevin Equation satisfies *microreversibility*, and therefore *detailed balance*. As is well known [8] and as we will show in a next section, this implies that the distribution of particles converges to the Boltzmann distribution.

1.10 Memory Kernel

As we mentioned at the beginning of this chapter, in some cases, the friction term is not represented by the instantaneous Stokes law, but it integrates the history of the system at previous times.

Under quite general conditions, the friction term can be represented as the convolution of a *memory kernel* with the velocity of the particle [9]. Such is the case for instance in the description of the motion of a particle in a visco-elastic fluid [10], or in the modeling of an effective degree of freedom, such as a reaction coordinate. Another example appears when studying the dynamics of a specific monomer in a polymer. One can then show that the friction term is non-local in time and has the form of a memory kernel [11]. A simple model for the interaction of a particle with a heat bath has been studied extensively. In this model, the heat bath is represented as a collection of harmonic oscillators (which may represent the vibrational modes of the solvent), coupled linearly to the coordinates of the studied particle [12]. One can integrate out the harmonic oscillator degrees of freedom. The result of this reduction of degrees of freedom is the generation of a history dependent friction term, as well as a random force acting on the particle.

We discuss only the case of the overdamped Langevin Equation. The generalized Langevin Equation (GLE) takes the form

$$\int_0^t d\tau \Gamma(t - \tau)\dot{x}(\tau) = F(x(t)) + \xi(t) \tag{1.136}$$

where $\Gamma(t)$ is the memory kernel, and as usual, $\xi(t)$ is a Gaussian noise, to be specified below. The upper bound of the integral above can be extended to $+\infty$, provided we impose the condition $\Gamma(\tau) = 0$ if $\tau < 0$. The memory kernel must be definite positive, so as to represent a dissipative term, and decreasing in time. Standard forms include exponential decay $e^{-\frac{t}{\tau}}$ or algebraic decay $t^{-\alpha}$. Obviously, the GLE is not Markovian, as the state of the system at time t depends on its history at all previous times $\tau < t$.

The random noise $\xi(t)$ should clearly have a vanishing average at all time

$$\langle \xi(t) \rangle = 0 \tag{1.137}$$

otherwise it would amount to applying an average force on the system. In addition, it can be shown that if one requires that the system satisfies detailed balance [9, 12], the correlation function of the noise must be given by

$$\langle \xi(t) \rangle = 0$$
$$\langle \xi(t)\xi(t') \rangle = k_B T \Gamma(t - t') \tag{1.138}$$

Note that one recovers the standard memory-less Langevin Equation if the noise correlation is given by

$$\Gamma(t - t') = 2\gamma \delta(t - t') \tag{1.139}$$

Indeed, the left hand-side of Eq. (1.136) becomes

$$\int_0^t d\tau \Gamma(t-\tau)\dot{x}(\tau) = 2\gamma \int_0^t d\tau \delta(t-\tau)\dot{x}(\tau)$$
$$= \gamma \dot{x}(t) \tag{1.140}$$

where we have used the fact that the τ integral is integrated only over $\tau < t$ and thus

$$\int_0^t d\tau \delta(t-\tau)\dot{x}(\tau) = \frac{1}{2}\dot{x}(t) \tag{1.141}$$

The general result is that in presence of a non-trivial memory kernel, in order to satisfy detailed balance, the noise must be Gaussian coloured, given by Eq. (1.138).

1.11 The Many Particle Case

We write explicitly the Langevin Equation for a 3-dimensional system of N particles interacting through a one-body potential V and a two-body potential v. The potential energy of the system reads

$$U(\{\vec{r}_i\}) = \sum_{i=1}^N V(\vec{r}_i) + \frac{1}{2}\sum_{i\neq j} v(\vec{r}_i - \vec{r}_j) \tag{1.142}$$

and the force is

$$\vec{F}_i = -\frac{\partial U}{\partial \vec{r}_i} = -\frac{\partial V(\vec{r}_i)}{\partial \vec{r}_i} - \sum_{j\neq i} \frac{\partial v(\vec{r}_i - \vec{r}_j)}{\partial \vec{r}_i} \tag{1.143}$$

The Langevin Equation becomes

$$\frac{d\vec{r}_i}{dt} = \frac{1}{\gamma}\vec{F}_i + \vec{\eta}_i \tag{1.144}$$

with

$$\langle \eta_i^\alpha(t)\rangle = 0$$
$$\langle \eta_i^\alpha(t)\eta_j^{\alpha'}(t')\rangle = 2D\delta_{ij}\delta_{\alpha\alpha'}\delta(t-t') \tag{1.145}$$

for $i, j = 1, \ldots, N$ and $\alpha, \alpha' = x, y, z$ are the spatial indices.

A similar equation can be written for the underdamped case (mass term), and one may use any proper discretization to solve these equations numerically,

References

1 Langevin, P. (1908). On the theory of Brownian motion, *C. R. Acad. Sci. (Paris)* 146: 530–533.

2 (a) Risken, H. (1989). *The Fokker–Planck Equation. Berlin, Heidelberg, New-York, London, Paris,* Tokyo: Springer-Verlag. (b) Van Kampen, N.G. (2007). *Stochastic Processes in Physics and Chemistry.* North-Holland. (c) Zwanzig, R. (2001). *Nonequilibrium Statistical Mechanics.* Oxford University Press.

3 Gardiner, C. (2009). *Stochastic Methods: A Handbook for the Natural and Social Sciences.* Springer Series in Synergetics, Berlin, Heidelberg: Springer-Verlag.

4 Landau, L.D. and Lifschitz, E.M. (1959). *Fluid Mechanics.* London: Pergamon Press.

5 Einstein, A. (1905). Über die von der molekularkinetischen Theorie der Wärme geforderte Bewegung von in ruhenden Flüssigkeiten suspendierten Teilchen, *Ann. Phys.* 322 (8): 549

6 Feller, W. (1971). *An Introduction to Probability Theory and Its Applications,* vol. II, 2e. New York, NY: Wiley.

7 Wong, E. and Zakai, M. (1965). On the convergence of ordinary integrals to stochastic integrals, *Ann. Math. Statist.* 36: 1560.

8 Chandler, D. (1987). *Introduction to Modern Statistical Mechanics,* New-York: Oxford University Press.

9 (a) Zwanzig, R. (1961). Memory effects in irreversible thermodynamics, *Phys. Rev.* 124: 983. (b) Mori, H. (1965). Transport, Collective Motion, and Brownian Motion, *Prog. Theor. Phys.* 33: 423. (c) Kubo, R. (1965). *Many-Body Theory, Part 1, Tokyo Summer Lectures in Theoretical Physics,* (ed. R. Kubo). Tokyo/New-York: Shokabo/Benjamin.

10 (a) Hess, W. and Klein, R. (1983). Generalized hydrodynamics of systems of Brownian particles, *Adv. Phys.* 32 (2): 173–283. (b) Mason, T.G. and Weitz, D.A. (1995). Optical measurements of frequency-dependent linear viscoelastic moduli of complex fluids. *Phy. Rev. Lett.* 74: 1250. (c) Chaikin, P.M. and Lubensky, T.C. (1995). *Principles of Condensed Matter Physics.* Cambridge University Press.

11 Sakaue, T., Walter, J.-C. and Carlon, E. et al. (2017). Non-Markovian dynamics of reaction coordinate in polymer folding, *Soft Matter* 13: 3174.

12 Zwanzig, R. (1973). Nonlinear generalized langevin equations, *J. Stat. Phys.* 9: 3.

2

The Fokker–Planck Equation

The Langevin equation is a stochastic differential equation (SDE) and the position and velocity of the particle are random variables. It is thus natural to define the *probability distribution function* or *p.d.f.* of the position of the particle at time t

$$P(x, t) = \langle \delta(x - x(t)) \rangle \tag{2.1}$$

for the case of the overdamped Langevin equation, and similarly, a probability distribution for the position and momentum of the particle

$$P(x, p, t) = \langle \delta(x - x(t))\delta(p - p(t)) \rangle \tag{2.2}$$

for the underdamped Langevin equation. As we will show below, this p.d.f. satisfies a differential equation, called the Smoluchowski or the Fokker–Planck equation (depending on the presence of the mass term in the Langevin Equation), and we refer the reader to the references [2] of chapter 1 for further developments.

2.1 The Chapman–Kolmogorov Equation

In this section, we restrict ourselves to the overdamped case, the underdamped case being a simple generalization. The Markovian property of the dynamics of the Langevin Equation allows to write the following equation

$$P(x, t | x_0, t_0) = \int dy P(x, t | y, \tau) P(y, \tau | x_0, t_0) \tag{2.3}$$

where $P(x, t | y, \tau)$ is the conditional probability for the particle to be at (x, t), given that it was at y at time $\tau < t$. This expresses the fact that in a Markov process, the probability to go from point (x_0, t_0) to point (x, t) is the sum of the products of the probability to go from (x_0, t_0) to any point y at time $\tau < t$ by the probability to go from (y, τ) to (x, t). We emphasize that the time τ is any time in the interval $[t_0, t]$. This equation is called the *Chapman–Kolmogorov*

Molecular Kinetics in Condensed Phases: Theory, Simulation, and Analysis,
First Edition. Ron Elber, Dmitrii E. Makarov and Henri Orland.
© 2020 John Wiley & Sons Ltd. Published 2020 by John Wiley & Sons Ltd.

equation [1]. If the initial point x_0 is itself distributed with a certain probability distribution $P_0(x_0) = P(x_0, t_0)$, then the probability to find the particle at point x at time t is given by

$$P(x, t) = \int dx_0 P(x, t | x_0, t_0) P(x_0, t_0) \tag{2.4}$$

Multiplying on the right both sides of Eq. (2.3) by $P_0(x_0)$ and integrating over x_0, the Chapman–Kolmogorov equation can be generalized to

$$
\begin{aligned}
P(x, t) &= \int dx_0 P(x, t | x_0, t_0) P_0(x_0) \\
&= \int dy P(x, t | y, \tau) \int dx_0 P(y, \tau | x_0, t_0) P_0(x_0) \\
&= \int dy P(x, t | y, \tau) P(y, \tau)
\end{aligned}
\tag{2.5}
$$

It is easily derived from the Bayes theorem for conditional probabilities [2]. Indeed, according to this theorem, we have

$$P(x, t | y, \tau) = \frac{\mathrm{Proba}(x, t \ \text{and} \ y, \tau)}{P(y, \tau)} \tag{2.6}$$

so that

$$\mathrm{Proba}(x, t \ \text{and} \ y, \tau) = P(x, t | y, \tau) P(y, \tau) \tag{2.7}$$

We have

$$P(x, t) = \int dy \ \mathrm{Proba}(x, t \ \text{and} \ y, \tau) \tag{2.8}$$

Indeed, integrating $\mathrm{Proba}(x, t \ \text{and} \ y, \tau)$ over y yields the probability to find the particle at x at time t. Therefore, integrating Eq. (2.7) over y yields Eq. (2.5). In plain words, this equation expresses the fact that for any time τ, $t_0 < \tau < t$, the probability for a particle to be at x at time t, is the sum of the products of the probability for the particle to be at any point y at time τ by the conditional probability for the particle to be at x at time t given that it was at y at time τ.

2.2 The Overdamped Case

2.2.1 Derivation of the Smoluchowski (Fokker–Planck) Equation using the Chapman–Kolmogorov Equation

In order to get a partial differential equation for P, it is useful to write the Chapman–Kolmogotov equation (2.5) for an infinitesimal time increment

$$P(x, t + dt) = \int dy P(x, t + dt | y, t) P(y, t) \tag{2.9}$$

Using the Itô scheme in its differential form, and using Eq. (1.130), we have

$$
P(x, t + dt | y, t) = \frac{1}{\sqrt{4\pi D dt}} e^{-\frac{\left(x - y - \frac{1}{\gamma} F(y) dt\right)^2}{4D dt}}
$$

$$
= \int \frac{dp}{2\pi} e^{-Dp^2 dt + ip\left(x - y - \frac{1}{\gamma} F(y) dt\right)} \tag{2.10}
$$

where we have used the Fourier representation of the Gaussian. Equation (2.9) becomes

$$
P(x, t + dt) = \int \frac{dp}{2\pi} \int dy\, e^{-Dp^2 dt + ip\left(x - y - \frac{1}{\gamma} F(y) dt\right)} P(y, t) \tag{2.11}
$$

and expanding to order 1 in dt, we obtain

$$
P(x, t + dt) = \int \frac{dp}{2\pi} \int dy \left(1 - Dp^2 dt - \frac{ip}{\gamma} F(y) dt\right) e^{ip(x-y)} P(y, t)
$$

$$
= \int \frac{dp}{2\pi} \int dy \left(1 + Ddt \frac{\partial^2}{\partial x^2} - \frac{dt}{\gamma} \frac{\partial}{\partial x} F(y)\right) e^{ip(x-y)} P(y, t)
$$

$$
= \int dy \left(1 + Ddt \frac{\partial^2}{\partial x^2} - \frac{dt}{\gamma} \frac{\partial}{\partial x} F(y)\right) \int \frac{dp}{2\pi} e^{ip(x-y)} P(y, t)
$$

$$
= \left(1 + Ddt \frac{\partial^2}{\partial x^2}\right) P(x, t) - \frac{dt}{\gamma} \frac{\partial}{\partial x} (F(x) P(x, t)) \tag{2.12}
$$

where we have used the identities

$$
\int \frac{dp}{2\pi} e^{ip(x-y)} = \delta(x - y) \tag{2.13}
$$

and

$$
\int dy \int \frac{dp}{2\pi} ipF(y) e^{ip(x-y)} = \int dy \int \frac{dp}{2\pi} \frac{\partial}{\partial x} F(y) e^{ip(x-y)} = \frac{\partial}{\partial x} F(x) \tag{2.14}
$$

$$
\int dy \int \frac{dp}{2\pi} p^2 F(y) e^{ip(x-y)} = -\int dy \int \frac{dp}{2\pi} \frac{\partial^2}{\partial x^2} F(y) e^{ip(x-y)} = -\frac{\partial^2}{\partial x^2} F(x) \tag{2.15}
$$

Equation (2.12) can be recast into the partial differential equation

$$
\frac{dP}{dt} = D \frac{\partial}{\partial x} \left(\frac{\partial P}{\partial x} - \beta F(x) P\right) \tag{2.16}
$$

or

$$
\frac{dP}{dt} = D \frac{\partial}{\partial x} \left(\frac{\partial P}{\partial x} + \beta \frac{\partial U(x)}{\partial x} P\right) \tag{2.17}
$$

if the force F is conservative, derived from the potential $U(x)$. This equation is a first order differential equation in time and should be supplemented by the initial condition

$$P(x, t_0) = P_0(x) \tag{2.18}$$

where $P_0(x)$ is the probability distribution of the particle at time $t = t_0$ For instance, for a particle localised at $x = x_0$ at time $t = t_0$, one would have $P_0(x) = \delta(x - x_0)$.

Equation (2.16) is a linear partial differential equation, which was derived by Smoluchowski [2, 3]. It is a special case of a more general equation, called the Fokker–Planck equation [2, 4] which is the general equation for the under-damped case. In the following chapters of this book, as in many texts, we will often refer to the Smoluchowski equation as the Fokker–Planck equation.

In the case of a conservative force, this equation can be written in the compact form

$$\frac{dP}{dt} = D \frac{\partial}{\partial x} \left(e^{-\beta U(x)} \frac{\partial}{\partial x} (e^{\beta U(x)} P) \right) \tag{2.19}$$

Being linear, Eq. (2.16) can be recast in the form

$$\frac{dP}{dt} = -L_{FP} P \tag{2.20}$$

where L_{FP} is a linear differential operator easily obtained from Eq. (2.16) or (2.19)

$$\begin{aligned} L_{FP} &= -D \left(\frac{\partial^2}{\partial x^2} - \beta \frac{\partial}{\partial x} (F(x) \, .) \right) \\ &= -D \left(\frac{\partial^2}{\partial x^2} - \beta F(x) \frac{\partial}{\partial x} - \beta \frac{\partial F(x)}{\partial x} \right) \end{aligned} \tag{2.21}$$

The operator L_{FP} is not *self-adjoint* (or symmetric), which means that its action on a function on the right is not identical to that on a function on the left, and we will study some of the consequences of this property later in this chapter.

In the case of N degrees of freedom x_i, with a diffusion constant D_i for each degree of freedom, this equation can be easily generalized to

$$\frac{dP(x_1, \dots, x_N, t)}{dt} = \sum_{i=1}^{N} D_i \frac{\partial}{\partial x_i} \left(\frac{\partial P}{\partial x_i} + \beta \frac{\partial U(x_1, \dots, x_N)}{\partial x_i} P \right) \tag{2.22}$$

Equations (2.16, 2.17) take the form of a *continuity equation*

$$\frac{dP}{dt} + \frac{\partial q}{\partial x} = 0 \tag{2.23}$$

with the current given by

$$q(x,t) = -D\left(\frac{\partial P}{\partial x} - \beta F(x)P\right)$$

$$= -D\left(\frac{\partial P}{\partial x} + \beta \frac{\partial U(x)}{\partial x}P\right) \tag{2.24}$$

The equilibrium solution to Eq. (2.23) is $q = 0$ which in the case of a conservative force, reads

$$\frac{\partial P_{eq}}{\partial x} = -\beta \frac{\partial U(x)}{\partial x}P_{eq} \tag{2.25}$$

The solution of this equation is the Boltzmann distribution

$$P_{eq}(x) = \frac{e^{-\beta U(x)}}{Z} \tag{2.26}$$

where Z is the partition function of the system. Note that the stationarity of the Boltzmann distribution could readily be seen by inserting $P_{eq}(x) = \frac{e^{-\beta U(x)}}{Z}$ in Eq. (2.19). We will show in subsection 2.3 that indeed, the probability distribution converges to the Boltzmann distribution at large times.

2.2.2 Alternative Derivation of the Smoluchowski (Fokker–Planck) Equation

It is instructive to see how simple stochastic calculus allows to retrieve the Fokker–Planck equation in a few lines. We assume that we use the Itô scheme of Eq. (1.60). We define $x(t + dt) = x(t) + dx(t)$. Expanding to 1st order in dt and to 2nd order in $dx(t)$ (since $dx(t)$ is of order $\sqrt{dt}$), we have

$$
\begin{aligned}
dP &= P(x, t + dt) - P(x, t) \\
&= \langle \delta(x - x(t + dt)) \rangle - \langle \delta(x - x(t)) \rangle \\
&= \langle \delta(x - x(t)) - dx(t)) \rangle - \langle \delta(x - x(t)) \rangle \\
&= -\left\langle dx(t)\frac{\partial}{\partial x}\delta(x - x(t)) \right\rangle + \frac{1}{2}\left\langle (dx(t))^2 \frac{\partial^2}{\partial x^2}\delta(x - x(t)) \right\rangle \\
&= -\frac{\partial}{\partial x}\left\langle \left(\frac{1}{\gamma}F(x(t))dt + dB(t)\right)\delta(x - x(t)) \right\rangle \\
&\quad + Ddt\frac{\partial^2}{\partial x^2}\langle \delta(x - x(t)) \rangle
\end{aligned} \tag{2.27}
$$

where we have used the identity (1.92).

Using the fact that the variable $x(t)$ at t depends only on the noise at times $t' < t$, Eq. (1.84), we have

$$\langle dB(t)\delta(x - x(t)) \rangle = 0 \tag{2.28}$$

so that Eq. (2.27) becomes

$$\frac{dP}{dt} = D\frac{\partial}{\partial x}\left(\frac{\partial P}{\partial x} - \beta F(x)P\right) \tag{2.29}$$

which is Eq. (2.16)

2.2.3 The Adjoint (or Reverse or Backward) Fokker–Planck Equation

The Chapman–Kolmogorov equation can be used to describe the backward evolution of the particle, from its final position to its initial one. Indeed, using Eq. (2.3), we may write

$$P(x, t|x_0, t_0) = \int dy P(x, t|y, t_0 + dt)P(y, t_0 + dt|x_0, t_0) \tag{2.30}$$

and expanding to first order in dt as in subsection 2.1.1, we obtain the so-called *adjoint* or *reverse* or *backward Fokker–Planck equation*

$$\frac{dP(x, t|y, \tau)}{d\tau} = -D\left(\frac{\partial^2 P(x, t|y, \tau)}{\partial y^2} + \beta F(y)\frac{\partial P(x, t|y, \tau)}{\partial y}\right) \tag{2.31}$$

This equation describes how the initial point is evolved, so that the particle reaches the target point x at time t. This equation is the adjoint (in the mathematical sense) of Eq. (2.20) and can be written as

$$\frac{dP(x, t|y, \tau)}{d\tau} = L_{FP}^{\dagger}P(x, t|y, \tau) \tag{2.32}$$

with

$$L_{FP}^{\dagger} = -D\left(\frac{\partial^2}{\partial y^2} + \beta F(y)\frac{\partial}{\partial y}\right) \tag{2.33}$$

This equation will be used to compute first passage times and rates, and also to stochastically sample transition paths (see sections 2.6 and 7.3).

2.3 The Underdamped Case

In the underdamped case, both the position and the momentum of the particle are random variables, and we must thus consider the p.d.f.

$$P(x, p, t) = \langle \delta(x - x(t))\delta(p - p(t))\rangle \tag{2.34}$$

Using the Langevin Eq. (1.64) and following the method used above in section 2.2.2, we compute the differential of this quantity to order dt, and obtain

$$
\begin{aligned}
dP &= P(x, p, t + dt) - P(x, p, t) \\
&= \langle \delta(x - x(t + dt))\delta(p - p(t + dt)) - \delta(x - x(t))\delta(p - p(t)) \rangle \\
&= \langle \delta(x - x(t) - dx(t))\delta(p - p(t) - dp(t)) - \delta(x - x(t))\delta(p - p(t)) \rangle \\
&= \left\langle \left(-dx(t)\frac{\partial}{\partial x} - dp(t)\frac{\partial}{\partial p} + \frac{1}{2}(dp(t))^2 \frac{\partial^2}{\partial p^2} \right) \delta(x - x(t))\delta(p - p(t)) \right\rangle \\
&= \left\langle \left(-\frac{p(t)}{m}dt\frac{\partial}{\partial x} - \frac{\partial}{\partial p}\left(-\frac{\gamma}{m}p(t)dt + Fdt + dB_1(t) \right) + \frac{1}{2}(dp)^2 \frac{\partial^2}{\partial p^2} \right) \right. \\
&\qquad \left. \times\, \delta(x - x(t))\delta(p - p(t)) \right\rangle
\end{aligned}
\tag{2.35}
$$

Using again the causality, we have

$$
\langle dB_1(t)\delta(x - x(t))\delta(p - p(t)) \rangle = 0
\tag{2.36}
$$

and the properly scaled Eq. (1.85)

$$
(dp)^2 = (dB_1(t))^2 = 2\gamma k_B T dt
\tag{2.37}
$$

we obtain the Fokker–Planck equation

$$
\frac{dP}{dt} = \gamma k_B T \frac{\partial^2 P}{\partial p^2} - \frac{p}{m}\frac{\partial P}{\partial x} + \frac{\partial}{\partial p}\left(\left(\frac{\gamma}{m}p - F \right) P \right)
\tag{2.38}
$$

As in the overdamped case, for a conservative force $F = -\frac{\partial U}{\partial x}$, we can check that the equilibrium distribution of Eq. (2.38) is the Boltzmann distribution

$$
P_{eq}(x, p) = \frac{e^{-\beta\left(\frac{p^2}{2m} + U(x) \right)}}{Z}
\tag{2.39}
$$

where Z is the partition function. Indeed, the r.h.s of Eq. (2.38) vanishes for $P = P_{eq}(x, p)$ above.

2.4 The Free Case

The case of a Brownian particle with no external force can be easily solved for both the over- and underdamped cases.

2.4.1 Overdamped Case

In that case, the Fokker–Planck equation (2.16) takes the form of the heat equation [2]

$$
\frac{dP(x, t)}{dt} = \frac{\partial^2 P}{\partial x^2}
\tag{2.40}
$$

with a specified boundary condition. The solution of this equation is well-known. It can be obtained by taking a Fourier transform of this equation (see [2]). For an initial condition where the particle is localized at x_i at time 0

$$P(x, 0) = \delta(x - x_i) \tag{2.41}$$

the solution is

$$P(x, t) = \frac{1}{\sqrt{4\pi Dt}} e^{-\frac{(x-x_i)^2}{4Dt}} \tag{2.42}$$

In the more general case where the initial condition is

$$P(x, 0) = P_0(x) \tag{2.43}$$

the solution is

$$P(x, t) = \frac{1}{\sqrt{4\pi Dt}} \int dy \, e^{-\frac{(x-y)^2}{4Dt}} P_0(y) \tag{2.44}$$

2.4.2 Underdamped Case

The Fokker–Planck equation (2.38) takes the form

$$\frac{dP}{dt} = \gamma k_B T \frac{\partial^2 P}{\partial p^2} - \frac{p}{m} \frac{\partial P}{\partial x} + \frac{\partial}{\partial p}\left(\frac{\gamma}{m} pP\right) \tag{2.45}$$

Assuming for example an initial condition where the particle is immobile at the origin

$$P(x, p, 0) = \delta(x)\delta(p) \tag{2.46}$$

and defining the Fourier transform

$$\tilde{P}(k, q, t) = \int dx \, dp \, e^{i(kx+pq)} P(x, p, t) \tag{2.47}$$

Equation (2.45) becomes

$$\frac{d\tilde{P}}{dt} = -\gamma k_B T q^2 \tilde{P} + \left(\frac{k - \gamma q}{m}\right) \frac{\partial \tilde{P}}{\partial q} \tag{2.48}$$

with

$$\tilde{P}(k, q, 0) = 1 \tag{2.49}$$

Defining

$$\Phi(k, q, t) = \ln \tilde{P}(k, q, t) \tag{2.50}$$

Eq. (2.48) becomes

$$\frac{d\Phi}{dt} = -\gamma k_B T q^2 + \left(\frac{k - \gamma q}{m}\right) \frac{\partial \Phi}{\partial q} \tag{2.51}$$

with the initial condition

$$\Phi(k, q, 0) = 0 \tag{2.52}$$

This equation can be easily solved by assuming a quadratic form for Φ

$$\Phi(k, q, t) = -\frac{1}{2} \begin{pmatrix} k & q \end{pmatrix} \begin{bmatrix} A(t) & B(t) \\ B(t) & C(t) \end{bmatrix} \begin{pmatrix} k \\ q \end{pmatrix}$$

$$= -\frac{1}{2} A(t) k^2 - B(t) kq - \frac{1}{2} C(t) q^2 \tag{2.53}$$

Substituting back in Eq. (2.51) we obtain the solution

$$A(t) = 2Dt - \frac{mk_B T}{\gamma^2} \left(3 - 4e^{-\frac{\gamma}{m}t} + e^{-\frac{2\gamma}{m}t} \right)$$

$$B(t) = \frac{mk_B T}{\gamma} \left(1 - e^{-\frac{\gamma}{m}t} \right)^2$$

$$C(t) = mk_B T \left(1 - e^{-\frac{2\gamma}{m}t} \right) \tag{2.54}$$

Taking the inverse Fourier transform of Eq. (2.47) we obtain

$$P(x, p, t) = \frac{1}{\mathcal{N}} \exp \left(-\frac{1}{2} \begin{pmatrix} x & p \end{pmatrix} \begin{bmatrix} A(t) & B(t) \\ B(t) & C(t) \end{bmatrix}^{-1} \begin{pmatrix} x \\ p \end{pmatrix} \right)$$

$$= \frac{1}{\mathcal{N}} e^{-\frac{C(t)x^2 - 2B(t)xp + A(t)p^2}{2\Delta}} \tag{2.55}$$

where $\Delta = A(t)C(t) - B^2(t)$ and $\mathcal{N} = 2\pi\sqrt{\Delta}$ is a normalization constant. This last equation (2.55) implies

$$\langle x(t) \rangle = 0$$
$$\langle x^2(t) \rangle = A(t)$$
$$\sim 2Dt \quad \text{when} \quad t \to \infty \tag{2.56}$$

and

$$\langle p(t) \rangle = 0$$
$$\langle p^2(t) \rangle = C(t)$$
$$\sim mk_B T \quad \text{when} \quad t \to \infty \tag{2.57}$$

We thus recover the usual diffusion law for the position of the particle and the equipartition theorem for the momentum of the particle at large time.

2.5 Averages and Observables

One of the main uses of the probability distribution P is that it allows to compute thermal averages as a function of time. For example, for the overdamped case, the average of a function $A(x(t))$ is easily expressed as

$$\langle A(x(t)) \rangle = \int dx A(x) P(x, t) \tag{2.58}$$

The situation is a bit more complicated for averages involving the velocity of the particle. Consider for example the average velocity $v(t)$ of the particle. We have

$$v(t) = \langle \dot{x}(t) \rangle \tag{2.59}$$

On the other hand, averaging the Langevin Equation (1.41) over the noise, we have

$$\langle \dot{x}(t) \rangle = \frac{1}{\gamma} \langle F(x(t)) \rangle \tag{2.60}$$

Therefore, we have

$$v(t) = \langle \dot{x}(t) \rangle = \frac{1}{\gamma} \int dx F(x) P(x, t) \tag{2.61}$$

Similarly, if we want to compute the average of the square of the velocity, we use the differential form of the Langevin Equation (1.60)

$$\langle (dx)^2 \rangle = \langle \left(\frac{F(x(t))}{\gamma} dt \right)^2 + (dB(t))^2 + 2 \left(\frac{F(x(t))}{\gamma} dt \right) dB(t) \rangle$$

$$= \langle \left(\frac{F(x(t))}{\gamma} dt \right)^2 \rangle + 2D dt + 2 \langle \left(\frac{F(x(t))}{\gamma} dt \right) dB(t) \rangle \tag{2.62}$$

The term $\langle \left(\frac{F(x(t))}{\gamma} dt \right) dB(t) \rangle$ in the r.h.s. vanishes because of the property Eq. (1.84), so that finally

$$\langle \left(\frac{dx}{dt} \right)^2 \rangle = \int dx \left(\frac{F(x)}{\gamma} \right)^2 P(x, t) + \frac{2D}{dt} \tag{2.63}$$

which shows that the average square velocity diverges in the continuous limit of the overdamped Brownian motion.

The case of the underdamped Langevin Equation is more straightforward. Indeed, the constituent equation

$$p = m\dot{x} \tag{2.64}$$

implies that for any observable $A(x(t), \dot{x}(t))$, we have

$$\langle A(x(t), \dot{x}(t)) \rangle = \int dx \int dp A(x, \frac{p}{m}) P(x, p, t) \tag{2.65}$$

References

1 Chapman, S. (1928). On the Brownian displacement and thermal diffusion of grains suspended in a non-uniform fluid, *Proc. Roy. Soc. London, Ser. A* 119: 34–54.

2 Kolmogorov, A.N. (1931). Uber die analytischen Methoden in der Wahrscheinlichkeitrechnung, *Math. Ann.* 104: 415–458.

3 Papoulis, A. and Pillai, S.U. (2002). *Probability, Random Variables, and Stochastic Processes, 4e*. New York: McGraw-Hill.

4 Feller, W. (1971). *An Introduction to Probability Theory and Its Applications, Vol. II, 2e*. New York, NY: Wiley.

5 Smoluchowski, M. (1906). Zur kinetischen theorie der brownschen molekularbewegung und der suspensionen, *Ann. Phys.* 21 (14): 756.

6 Fokker, A.D. (1914). Die mittlere Energie rotierender elektrischer Dipole im Strahlungsfeld, *Ann. Phys.* 348 (4. Folge 43): 810.

7 Planck, M. (1917). Über einen Satz der statistischen Dynamik und seine Erweiterung in der Quantentheorie, *Sitzungsberichte der Preussischen Akademie der Wissenschaften zu Berlin* 24: 324.

3

The Schrödinger Representation

As was shown in Eq. (2.20), the Fokker–Planck equation (2.16) can be written in the form

$$\frac{dP(x,t)}{dt} = -L_{FP}P(x,t) \tag{3.1}$$

where L is a differential linear operator. This operator is not self-adjoint, that is, its action on a right vector P is not identical to that on a left vector. One of the consequences of this absence of symmetry is that the left eigenvectors of the operator L are not identical to the right eigenvectors. However, the left and right eigenvectors corresponding to a given eigenvalue are orthogonal. In the conservative case, where the force can be derived from a potential $U(x)$, we will show that there is a simple transformation which transforms the operator L into a self-adjoint operator. Self-adjoint operators have very interesting mathematical properties, in particular, their left and right eigenvectors are identical, and they provide an orthonormal basis of the vector space. In the following chapters, we will indeed mostly consider systems with conservative forces, in the over-damped limit. In that case, we will show that the Smoluchowski equation (2.17) can be recast in the form of an imaginary time Schrödinger equation, with a self-adjoint operator.

3.1 The Schrödinger Equation

In the case when the potential $U(x)$ is not singular, we can define a new function $\Psi(x, t)$ by

$$P(x, t) = e^{-\frac{\beta U(x)}{2}} \Psi(x, t) \tag{3.2}$$

Molecular Kinetics in Condensed Phases: Theory, Simulation, and Analysis,
First Edition. Ron Elber, Dmitrii E. Makarov and Henri Orland.
© 2020 John Wiley & Sons Ltd. Published 2020 by John Wiley & Sons Ltd.

The derivatives of P can be easily calculated

$$\frac{dP}{dt} = e^{-\frac{\beta U(x)}{2}} \frac{d\Psi}{dt}$$

$$\frac{\partial P}{\partial x} = -\frac{\beta}{2} \frac{\partial U}{\partial x} e^{-\frac{\beta U(x)}{2}} \Psi + e^{-\frac{\beta U(x)}{2}} \frac{\partial \Psi}{\partial x}$$

$$\frac{\partial^2 P}{\partial x^2} = \left(\left(\frac{\beta}{2} \frac{\partial U}{\partial x} \right)^2 - \frac{\beta}{2} \frac{\partial^2 U}{\partial x^2} \right) e^{-\frac{\beta U(x)}{2}} \Psi$$

$$- \beta \frac{\partial U}{\partial x} e^{-\frac{\beta U(x)}{2}} \frac{\partial \Psi}{\partial x} + e^{-\frac{\beta U(x)}{2}} \frac{\partial^2 \Psi}{\partial x^2} \tag{3.3}$$

Making the replacement (3.2) in Eq. (2.17), we obtain

$$\frac{d\Psi}{dt} = D \left(\frac{\partial^2 \Psi}{\partial x^2} - V(x)\Psi \right) \tag{3.4}$$

with

$$V(x) = \left(\frac{\beta}{2} \frac{\partial U}{\partial x} \right)^2 - \frac{\beta}{2} \frac{\partial^2 U}{\partial x^2} \tag{3.5}$$

Equation (3.4) has exactly the form of a Schrödinger equation in imaginary time [1]

$$\frac{d\Psi}{dt} = -H\Psi \tag{3.6}$$

where H is a quantum Hamiltonian given by

$$H = D \left(-\frac{\partial^2}{\partial x^2} + V(x) \right) \tag{3.7}$$

If the initial condition is $P(x, 0) = P_0(x)$, from Eq. (3.2), this translates on Ψ as

$$\Psi(x, 0) = e^{\frac{\beta U(x)}{2}} P_0(x) \tag{3.8}$$

As was mentioned in the beginning of this section, the Hamiltonian H is obviously self-adjoint, and therefore we will be able to use its orthonormal basis of eigenvectors to expand the probability distribution $P(x, t)$.

In the case of multiple degrees of freedom, the Schrödinger equation still has the form of Eq. (3.6), but with Hamiltonian

$$H = -D \sum_{i=1}^{N} \frac{\partial^2}{\partial x_i^2} + DV(x_1, \ldots, x_N) \tag{3.9}$$

and

$$V(x_1, \ldots, x_N) = \sum_{i=1}^{N} \left(\left(\frac{\beta}{2} \frac{\partial U}{\partial x_i} \right)^2 - \frac{\beta}{2} \frac{\partial^2 U}{\partial x_i^2} \right) \tag{3.10}$$

As we will study the Schrödinger equation (3.6), it is convenient to use the standard Dirac notation of quantum mechanics [1]. The state vector of the particle is denoted as $|\Psi(t)\rangle$, and with this notation, the wavefunction $\Psi(x,t)$ can be expressed as

$$\Psi(x,t) = \langle x|\Psi(t)\rangle \tag{3.11}$$

The solution of Eq. (3.6) can be formally written as

$$|\Psi(t)\rangle = e^{-Ht}|\Psi(0)\rangle \tag{3.12}$$

where the operator e^{-Ht} is the *evolution operator* for the particle. Similarly, if we consider the probability distribution $P(x,t|x_i,0)$, which is the p.d.f. $P(x,t)$ with initial condition $P(x,0) = \delta(x - x_i)$, we have according to Eqs. (3.8) and (3.12)

$$\Psi(x,t) = e^{\frac{\beta U(x_i)}{2}} \langle x|e^{-Ht}|x_i\rangle$$

which using Eq. (3.2) implies

$$P(x,t|x_i,0) = e^{-\frac{\beta(U(x)-U(x_i))}{2}} \langle x|e^{-Ht}|x_i\rangle \tag{3.13}$$

This fundamental relation will often be used in the following chapters of the book.

3.2 Spectral Representation

The Hamiltonian H being Hermitian, its eigenvalues are real and its eigenstates $|\Psi_n\rangle$, $n = 0, \ldots \infty$ with wavefunction $\Psi_n(x)$, form an orthonormal basis of the Hilbert space

$$H|\Psi_n\rangle = E_n|\Psi_n\rangle \tag{3.14}$$

We order the eigenvalues E_n according to their index n

$$E_0 < E_1 < \ldots < E_n < \ldots$$

and normalize the eigenstates

$$\langle \Psi_m|\Psi_n\rangle = \int dx \Psi_m(x)\Psi_n(x) = \delta_{mn}$$

Note that the eigenfunctions $\Psi_n(x)$ of H can be taken as real since the Hamiltonian H is real.

We may expand the evolution operator as

$$e^{-Ht} = \sum_{n=0}^{\infty} e^{-E_n t}|\Psi_n\rangle\langle \Psi_n| \tag{3.15}$$

Using Eq. (3.12) we have

$$\Psi(x, t) = e^{-Ht}\Psi(x, 0) = \sum_{n=0}^{\infty} e^{-E_n t} c_n \Psi_n(x) \tag{3.16}$$

where

$$c_n = \langle \Psi_n | \Psi(0) \rangle = \int dx \Psi_n(x)\Psi(x, 0) \tag{3.17}$$

3.3 Ground State and Convergence to the Boltzmann Distribution

Consider the normalized wavefunction

$$\Phi(x) = \frac{e^{-\frac{\beta U(x)}{2}}}{\sqrt{Z}} \tag{3.18}$$

where

$$Z = \int dx \, e^{-\beta U(x)} \tag{3.19}$$

By computing the derivatives of Φ

$$\frac{\partial \Phi}{\partial x} = -\frac{\beta}{2}\frac{\partial U}{\partial x}\Phi$$

$$\frac{\partial \Phi}{\partial x} = \left(\left(\frac{\beta}{2}\frac{\partial U}{\partial x} \right)^2 - \frac{\beta}{2}\frac{\partial^2 U}{\partial x^2} \right)\Phi$$

and using Eqs. (3.5) and (3.7), we readily see that

$$H|\Phi\rangle = 0 \tag{3.20}$$

There is a theorem in quantum mechanics which states that the ground-state wave function of any (regular enough) system is non-negative and is not degenerate. This statement is a special case of the Perron-Frobenius theorem [2], applied to the continuous case, which will be used again in section 4.2. Equation (3.20) shows that $|\Phi\rangle$ is an eigenstate of H with 0 eigenvalue and Eq. (3.18) shows that it is positive and has no zero. We conclude that $|\Phi\rangle$ is the ground state of H with eigenvalue $E_0 = 0$. It follows that $0 = E_0 < E_1 < \ldots < E_n < \ldots$. For large time, we can reduce the expansion (3.16) to the first excited state

$$\Psi(x, t) = c_0 \Psi_0(x) + e^{-E_1 t} c_1 \Psi_1(x) + \ldots \tag{3.21}$$

since higher order terms of the expansion are exponentially negligible. This expansion is valid for $t \gg \frac{1}{E_1}$. Equation (3.21) shows that the *relaxation time*

τ_r of the system is equal to $\frac{1}{E_1}$

$$\tau_r = \frac{1}{E_1} \tag{3.22}$$

where E_1 is the smallest non-zero positive eigenvalue of the Hamiltonian.
When $t \to \infty$, the system converges to

$$\Psi(x, t) = c_0 \Psi_0(x) \tag{3.23}$$

i.e.

$$\Psi(x, \infty) = c_0 \frac{e^{-\frac{\beta U(x)}{2}}}{\sqrt{Z}} \tag{3.24}$$

Going back to the probability distribution P, using the relation (3.2), we have

$$P(x, \infty) = c_0 \frac{e^{-\beta U(x)}}{\sqrt{Z}} \tag{3.25}$$

which after normalization becomes

$$P(x, \infty) = P_{eq}(x) = \frac{e^{-\beta U(x)}}{Z} \tag{3.26}$$

where P_{eq} denotes the equilibrium Boltzmann distribution. This last equation proves that the probability distribution converges to the Boltzmann distribution at large time

$$P(x, t) = \frac{e^{-\beta U(x)}}{Z} + e^{-E_1 t}\, e^{-\frac{\beta U(x)}{2}} c_1 \Psi_1(x) + \dots \tag{3.27}$$

To conclude this part, let us again emphasize that the transformation from Fokker–Planck to Schrödinger equation is possible only for conservative forces. In that case, it can be proven that the system will converge to thermal equilibrium. For non-conservative systems, there is no such convergence theorem, and the system may exhibit various types of long time behavior (stationary states, cycles, etc.).

As an example of application of this powerful method, we now show how this method can be applied to study the motion of a particle in a harmonic potential. As we have seen, a particle in a harmonic oscillator is described by an OU process. Assuming that the potential is

$$U(x) = \frac{K}{2} x^2 \tag{3.28}$$

the effective potential Eq. (3.5) is given by

$$V(x) = \frac{\beta^2 K^2}{4} x^2 - \frac{\beta K}{2} \tag{3.29}$$

and the "quantum" Hamiltonian is

$$H = -D\frac{d^2}{dx^2} + \frac{D\beta^2 K^2}{4}x^2 - \frac{D\beta K}{2} \tag{3.30}$$

The eigenstates and eigenvalues of the quantum harmonic oscillator are known [1]. For the standard quantum harmonic oscillator Hamiltonian

$$H_{HO} = -\frac{\hbar^2}{2m}\frac{d^2}{dx^2} + \frac{1}{2}m\omega^2 x^2 \tag{3.31}$$

the spectrum is given by $E_n^{(HO)} = \left(n + \frac{1}{2}\right)\hbar\omega$, with $n = 0, 1, \dots$ [1]. Adapting these results to the specific form (3.30) of the Hamiltonian, we make the correspondence

$$D = \frac{\hbar^2}{2m}$$
$$m\omega^2 = \frac{D\beta^2 K^2}{2} \tag{3.32}$$

so that

$$\frac{m\omega}{\hbar} = \frac{\beta K}{2} \tag{3.33}$$

The eigenvalues are given by

$$E_n = nD\beta K \tag{3.34}$$

with $n = 0, 1, \dots$ and the eigenfunctions by

$$\Psi_n(x) = \frac{1}{\sqrt{2^n n!}}\left(\frac{\beta K}{2\pi}\right)^{1/4} H_n\left(\frac{\beta K}{2}x\right)e^{-\frac{\beta K}{2}x^2} \tag{3.35}$$

where the $H_n(x)$ denote the Hermite polynomial defined by

$$H_n(x) = e^{\frac{x^2}{2}}\left(x - \frac{d}{dx}\right)^n e^{-\frac{x^2}{2}} \tag{3.36}$$

It follows that the ground state energy is $E_0 = 0$ and the relaxation time to equilibrium is given by the first eigenvalue

$$\tau_r = \frac{1}{E_1} = \frac{k_B T}{KD} \tag{3.37}$$

in agreement with Eq. (1.58). Similarly, the ground state of the Hamiltonian (3.31) is obtained by taking $n = 0$ in Eqs. (3.35) and (3.36). It is a Gaussian wave function, and consequently the equilibrium distribution of the system is given by the Boltzmann distribution

$$P_{eq}(x) = \sqrt{\frac{K\beta}{2\pi}}e^{-\frac{K\beta}{2}x^2} \tag{3.38}$$

References

1 Cohen-Tannoudji, C., Diu, B., and Laloe, F. (1977). *Quantum Mechanics*, vol. 1. Wiley. Feynman, R.P., Leighton, R.B., and Sands, M. (1971). *The Feynman Lectures on Physics*, vol. 3. Addison Wesley.
2 (a) Meyer, C. (2000). *Matrix Analysis and Applied Linear Algebra*, SIAM edition. (b) Winkler, G. (1995). *The Perron-Frobenius Theorem*, Springer-Verlag, Berlin Heidelberg.

4

Discrete Systems: The Master Equation and Kinetic Monte Carlo

4.1 The Master Equation

The Langevin equation is a stochastic differential equation, which assumes that space is continuous with a distance defined on it. In many cases, in physics, chemistry, biology, ecology,..., the states of a system are not specified by continuous variables but rather by discrete ones, possibly in infinite number. Such is the case of spin systems, or quantum atoms or molecules, etc. It is also the case in the study of the kinetics of protein folding, in the so-called *Markov State Model*, in which the trajectories are coarse-grained according to certain procedures [1]. Note that in addition to the states, the time may also be discrete. Other types of discretization of the phase space, generically called "celling" approaches are discussed in chapter 13 and 14 of the book.

For a thorough mathematical review of Markov processes, we refer the reader to [2].

If the set of states is discrete, it may be interesting to define the equivalent of a trajectory like the Langevin trajectory of a particle in continuous space, as well as the equivalent of a Fokker–Planck equation for the probability of the system to be in a specific state. A stochastic discrete trajectory would be a sequence of states that the system goes through in time. After giving a definition of Markov chains, we will show how to construct samples of trajectories using the *Kinetic Monte Carlo* method in subsection 4.3.

We first briefly describe *discrete-time Markov chains* and then *continuous-time Markov chains*, also called *Markov processes*. In the following, we will denote by $\mathcal{S} = \{a, b, ...\}$ the set of states of the system. The current state of the system at step n is denoted by s_n for discrete time and $s(t)$ for continuous time t.

4.1.1 Discrete-Time Markov Chains

By definition, a discrete-time Markov chain is a stochastic process in which the probability to find the system in state s_{n+1} at (discrete) time t_{n+1} depends solely

Molecular Kinetics in Condensed Phases: Theory, Simulation, and Analysis,
First Edition. Ron Elber, Dmitrii E. Makarov and Henri Orland.
© 2020 John Wiley & Sons Ltd. Published 2020 by John Wiley & Sons Ltd.

on the state s_n of the system at time t_n

$$P(s_{n+1}|s_n, s_{n-1}, \ldots, s_0) = P(s_{n+1}|s_n)$$
$$= W_{s_{n+1}s_n} \tag{4.1}$$

where $P(s_{n+1}|s_n, s_{n-1}, \ldots, s_0)$ denotes the probability for the system to be in state s_{n+1} at step $n + 1$, given that it went thought states s_0 at step 0, s_1 at step 1, ..., s_n at step n. Similarly, $P(s_{n+1}|s_n)$ is the probability for the system to be in state s_{n+1} at step $n + 1$, given that it was in state s_n at step n. This can be rephrased as follows: the system having gone through the sequence of states $\{s_0, s_1, \ldots, s_n\}$ until time t_n, the probability to find it in state s_{n+1} at time t_{n+1} depends only on the last visited state s_n at time t_n.

The matrix W_{ij} is called the *one-step transition probability matrix*. It is the conditional probability that the system being in state j at a given time, will jump to state i at the next time. Since it is a one step probability it has the obvious property that

- All the entries are non-negative: $W_{ij} \geq 0$
- Since when leaving j, the system must move to one state $i \in \mathcal{S}$, we have

$$\sum_{i \in \mathcal{S}} W_{ij} = 1 \tag{4.2}$$

A matrix W with these properties is called a *stochastic matrix*. An obvious property of stochastic matrices is that they have an eigenvalue equal to 1, corresponding to the left eigenvector $X_i = 1$ for all i. Indeed, from Eq. (4.2), we have

$$\sum_{i \in \mathcal{S}} X_i W_{ij} = X_j \tag{4.3}$$

with $X_i = 1$ for all i. Under certain additional conditions such as the irreducibility of the matrix[1], the *Perron-Frobenius* theorem [2] states that the maximum eigenvalue of the matrix W is unique and non-degenerate, and in addition, its left and right maximal eigenvectors are the only ones that have all of their components positive. So if the stochastic matrix W is irreducible, since $X_i = 1$ is a left eigenvector with all components positive and eigenvalue 1, it follows that 1 is the maximal eigenvalue of W and thus there exists a corresponding right eigenvector with all components positive, and this vector is an invariant distribution π_i of the discrete Markov chain, that is

$$\sum_{j \in \mathcal{S}} W_{ij} \pi_j = \pi_i \tag{4.4}$$

---

1 A matrix A is irreducible if for every pair of indices i and j, there exists a natural number m such that $(A^m)_{ij}$ is not equal to 0. In other words, for any pair of states i and j, there is a finite series of steps which connects them.

In the case when all the entries of the matrix are strictly positive, the Perron-Frobenius theorem implies that there is a finite gap between the two largest eigenvalues and the system does converge to the invariant distribution from any initial condition. We will further develop this point in the case of continuous Markov processes with detailed balance in subsection 4.2.

We can easily write a Chapman–Kolmogorov equation analogous to (2.3)

$$P(s_n|s_0) = \sum_{s_p} P(s_n|s_p)P(s_p|s_0) \tag{4.5}$$

for any p, $0 < p < n$. This property can be rephrased in the following way: the probability for the system starting in state s_0 at time 0 to be at s_n at time t_n is equal to the sum of the products of the probability of the system to start at s_0 and to be in any state s_p at time t_p by the probability for the system starting at s_p at time t_p to be at s_n at time t_n.

Choosing $p = n - 1$, this equation becomes

$$P(s_{n+1}|s_0) = \sum_{s_n} P(s_{n+1}|s_n)P(s_n|s_0)$$

$$= \sum_{s_n} W_{s_{n+1}s_n} P(s_n|s_0) \tag{4.6}$$

which is the constitutive equation for a discrete Markov chain. One can see from Eqs. (4.4) and (4.6) that if $P(s_n|s_0) = \pi_n$ then $P(s_{n+1}|s_0) = \pi_{n+1}$ which proves that $\{\pi_n\}$ is the invariant distribution of the Markov chain.

Iterating this equation, we obtain

$$P(s_n|s_0) = (W^n)_{s_n s_0} \tag{4.7}$$

where W^n denotes the nth power of the matrix W. When $n \to \infty$, we see that the behaviour of W^n is determined by the largest eigenvalue of W, which, as we saw above, is equal to 1. However, in case the matrix W is not irreducible, this does not necessarily imply that the system will converge to the invariant distribution π_n. We will discuss this point in more details in subsection 4.2.

4.1.2 Continuous-Time Markov Chains, Markov Processes

We now turn to the continuous time case. Continuous-time Markov chains are also called *Markov processes*.

We assume that the system in a given state a can randomly make a transition to a state b with a certain rate k_{ba} which will be defined below. The system is entirely specified by the $P_a(t)$ for all a, the probability for the system to be in state a at time t, and the evolution of the system is assumed to be governed by a so-called *master equation*. A continuous master equation is a set of linear equations which express the bookkeeping of probabilities, when the system evolves through stochastic jumps between states with a given rate matrix. The master equation is just the continuous time equivalent of the Chapman–Kolmogorov Eq. (4.6). The general form of the linear Markovian master equation is thus

$$\frac{dP_a(t)}{dt} = \sum_b (k_{ab} P_b(t) - k_{ba} P_a(t)) \tag{4.8}$$

or in differential form

$$dP_a(t) = P_a(t + dt) - P_a(t)$$
$$= \sum_b (k_{ab} P_b(t) - k_{ba} P_a(t)) dt \tag{4.9}$$

The matrix $k = \{k_{ab}\}$ is the rate matrix. All elements k_{ab} are positive (or zero), except possibly the diagonal terms k_{aa} which are obviously irrelevant (since they cancel out from the equation), and are equal to the transition rates from state b to state a. By definition of the rates, $k_{ab} dt$ is the probability for the system being in state b to make a transition to state a during time dt. It is thus positive as stated above. The physical interpretation of this equation is very simple: the first term of the r.h.s. of (4.9) represents the increase of the probability for the system to be in state a during time dt, due to the transitions from systems in any state b, whereas the second term represents its decrease due to the transitions from a to any state b during time dt. Using the notation

$$\Gamma_a = \sum_{b \neq a} k_{ba} \tag{4.10}$$

we see that the quantity $\Gamma_a dt$ is the probability for the system in a to jump to any other state b during time dt.

In the case of continuous space, as discussed in the previous sections, the space of states $\{a\}$ is the real Euclidean space and the transition rates $\{k_{ab}\}$ connect neighboring points in space $k(x, x + dx)$. By a proper choice of these transition rates, one can show that the continuous limit of the master equation (4.8) is the Fokker–Planck equation (2.16).

Equation (4.8) conserves the total probability. Indeed, summing this equation over all states a implies

$$\sum_a \frac{dP_a(t)}{dt} = \sum_{ab} (k_{ab} P_b(t) - k_{ba} P_a(t))$$
$$= \sum_{ab} k_{ab} P_b(t) - \sum_{ab} k_{ba} P_a(t)$$
$$= 0 \tag{4.11}$$

Thus $\sum_a P_a(t)$ is a constant in time, equal to 1.

The master equation can be written in differential matrix form as

$$P_a(t + dt) = \sum_b W_{ab} P_b(t) \tag{4.12}$$

with

$$W_{ab} = k_{ab}dt \text{ if } b \neq a$$
$$W_{aa} = 1 - \Gamma_a dt \tag{4.13}$$

This form is similar to the form (4.6) for the discrete case. The relation (4.12) allows to compute the probability at any time. Indeed if $t = ndt$, we have

$$P_a(t) = P_a(ndt) = \sum_b (W^n)_{ab} P_b(0) \tag{4.14}$$

where W^n is the nth power of the matrix W. The behaviour of the probabilities $P_a(t)$ at large time is determined by the largest eigenvalue and eigenvector of W. We will not discuss the general case here, but will focus on the case of detailed balance in subsection 4.2.

From the definition of the matrix W, we have

$$\sum_a W_{ab} = 1 \tag{4.15}$$

The matrix elements of k are all positive or zero and thus for small enough dt, the matrix W is also non-negative. From the property (4.15), W is a stochastic matrix. Therefore as we saw above, its largest eigenvalue is 1, and there exists an invariant distribution, to which the system converges at large time if the matrix W is irreducible.

The matrix W_{ab} is not symmetric, and therefore its left and right eigenvectors are not identical. However, as we will show later, in the case when there is *detailed balance* (equivalent to the case of a conservative force for the Fokker–Planck equation), one can bring it to a symmetric form by a simple transformation, and show that the system does converge to thermal equilibrium at large times.

4.2 Detailed Balance

A sufficient condition for the existence of a stationary distribution is the *detailed balance condition*, which can be expressed as

$$\frac{k_{ab}}{k_{ba}} = \frac{P_a^{(E)}}{P_b^{(E)}} \text{ for any pair of states } (a, b) \tag{4.16}$$

where $P_a^{(E)}$ is some equilibrium distribution. In the following, we will restrict the equilibrium distribution to be the Boltzmann distribution

$$\frac{k_{ab}}{k_{ba}} = e^{-\beta(E_a - E_b)} \text{ for any pair of states } (a, b) \tag{4.17}$$

where E_a is the energy of state a. An obvious consequence of detailed balance is that all transition rates are strictly positive and non-zero. Therefore, there are no absorbing states (sinks). If condition (4.17) is satisfied, it is easily seen that the Boltzmann distribution

$$P_a^{(E)} = \frac{e^{-\beta E_a}}{Z} \tag{4.18}$$

with

$$Z = \sum_a e^{-\beta E_a} \tag{4.19}$$

is a stationary solution of the master equation (4.8). Indeed, using the Boltzmann distribution (4.18), we have

$$k_{ab} P_b^{(E)} = k_{ba} P_a^{(E)} \tag{4.20}$$

for any pair (a, b), and thus Eq. (4.8) is satisfied with $\frac{dP_a^{(E)}}{dt} = 0$. However this does not prove that starting from any configuration, the system will converge to the Boltzmann distribution at large time. This will be proved in section (4.2.5)

Very often, in practical cases, the dynamics in state space results from a coarse-graining and the transition rates k_{ab} are not known exactly. One then resorts to some phenomenological approach to choose these rates, with the constraint of satisfying detailed balance. We list here several often used phenomenological forms for the transition rates.

4.2.1 Final State Only

$$k_{ab} = C e^{-\beta E_a} \tag{4.21}$$

i.e. the rates depend only on the final state a and C is some constant. Obviously, this form satisfies detailed balance. The master equation takes the form

$$\frac{dP_a(t)}{dt} = \sum_b (k_{ab} P_b(t) - k_{ba} P_a(t)) = C e^{-\beta E_a} - CZ P_a(t) \tag{4.22}$$

where $Z = \sum_a e^{-\beta E_a}$

The solution of this equation is

$$P_a(t) = \frac{e^{-\beta E_a}}{Z}(1 - e^{-CZt}) + P_a(0)e^{-CZt} \tag{4.23}$$

As expected, this probability converges to the Boltzmann distribution at large time.

4.2.2 Initial State Only

$$k_{ab} = C e^{\beta E_b} \tag{4.24}$$

i.e. the rates depend only on the initial state b and C is a constant. Obviously, this form satisfies detailed balance. It is used in modeling the diffusion of a system in random traps.

4.2.3 Initial and Final State

$$k_{ab} = C^{-\frac{\beta}{2}(E_a - E_b)} \tag{4.25}$$

i.e. the rates depend both on the initial and final states and C is a constant. More precisely, the rate is exponential in the barrier height $E_a - E_b$. Obviously, this form satisfies detailed balance. It is often used in the modeling of thermally activated barrier crossing.

4.2.4 Metropolis Scheme

The Metropolis scheme was developed for Monte Carlo calculations, but it can also be used to define the transition rates in a master equation. In the Metropolis scheme, we have

- If $E_a < E_b$, $k_{ab} = C$
- If $E_a > E_b$, $k_{ab} = Ce^{-\beta(E_a - E_b)}$

We see that if $E_a > E_b$, $k_{ab} = Ce^{-\beta(E_a - E_b)}$ and $k_{ba} = C$, so that the ratio of the k's satisfy the detailed balance equation (4.17). The same holds true for $E_a < E_b$.

4.2.5 Symmetrization

As we have seen, the matrix A is not symmetric, but by analogy with the Langevin case in section 3, we can make a change of function which renders the operator A of Eq. (4.12) symmetric. Following Eq. (3.2), we define $Q_a(t)$ by

$$P_a(t) = e^{-\frac{\beta}{2}E_a} Q_a(t) \tag{4.26}$$

then Eq. (4.12) becomes

$$\frac{dQ_a}{dt} = -\sum_b H_{ab} Q_b \tag{4.27}$$

with

$$H_{ab} = e^{\frac{\beta}{2}(E_a - E_b)} A_{ab}$$
$$H_{ab} = -e^{\frac{\beta}{2}(E_a - E_b)} k_{ab} \text{ if } a \neq b \tag{4.28}$$

$$H_{aa} = \sum_{b \neq a} k_{ba} = \Gamma_a \tag{4.29}$$

With this definition, we see that the operator H is symmetric. Indeed, we have

$$\frac{H_{ab}}{H_{ba}} = e^{\beta(E_a - E_b)} \frac{k_{ab}}{k_{ba}}$$
$$= 1 \tag{4.30}$$

due to Eq. (4.17).

We use the notation H_{ab} for the symmetrized operator to emphasize the analogy with the Hamiltonian of the Schrödinger equation (3.6) used to render the Fokker–Planck operator adjoint.

The matrix H can be written in a simpler form as

$$H_{ab} = \delta_{ab} \sum_c k_{ca} - e^{\frac{\beta}{2}(E_a - E_b)} k_{ab}$$

$$= \delta_{ab} \Gamma_a - e^{\frac{\beta}{2}(E_a - E_b)} k_{ab} \tag{4.31}$$

Define the maximum of all the Γ_a

$$\Gamma = \sup_a \Gamma_a \tag{4.32}$$

and the matrix M by

$$M_{ab} = H_{ab} - \Gamma \delta_{ab} = -e^{\frac{\beta}{2}(E_a - E_b)} k_{ab} - (\Gamma - \Gamma_a) \delta_{ab} \tag{4.33}$$

or in matrix form $M = H - \Gamma \mathbb{1}$ where $\mathbb{1}$ denotes the unity matrix. The matrix M is real symmetric, and all its elements are negative. It can be diagonalized and all its eigenvalues μ_n are real. The eigenvalues λ_n of the matrix $H = M + \Gamma \mathbb{1}$ are related to the eigenvalues μ_n of M by the relation $\lambda_n = \mu_n + \Gamma$. In particular, defining

$$Q_a^{(E)} = C e^{-\frac{\beta}{2} E_a} \tag{4.34}$$

where C is a normalization constant, we have

$$\sum_b M_{ab} Q_b^{(E)} = -\sum_b e^{\frac{\beta}{2}(E_a - E_b)} k_{ab} e^{-\frac{\beta}{2} E_b} - (\Gamma - \Gamma_a) e^{-\frac{\beta}{2} E_a}$$

$$= -e^{\frac{\beta}{2} E_a} \sum_b k_{ba} e^{-\beta E_a} - (\Gamma - \Gamma_a) e^{-\frac{\beta}{2} E_a}$$

$$= -\Gamma Q_a^{(E)} \tag{4.35}$$

where we have used the detailed balance identity $k_{ab} e^{-\beta E_b} = k_{ba} e^{-\beta E_a}$ from Eq. (4.17). The vector $Q_a^{(E)}$ is an eigenvector of M with eigenvalue $-\Gamma$. We may apply the Perron-Frobenius theorem to the positive matrix $-M$. According to this theorem, the above equation shows that $Q_a^{(E)}$ is an eigenvector of $-M$ with strictly positive components, therefore it is the maximal eigenvector of $-M$, with eigenvalue $-\mu_0 = \Gamma$. In addition, this eigenvalue is non-degenerate. It follows that $Q_a^{(E)}$ is the eigenvector of H with the smallest eigenvalue $\lambda_0 = 0$ and thus all eigenvalues of H are positive

$$0 = \lambda_0 < \lambda_1 < \ldots < \lambda_n < \ldots \tag{4.36}$$

Furthermore, there is a finite gap λ_1 between the ground state of H and its next eigenvalue. The fact that $Q_a^{(E)}$ is a 0 eigenvector of H can be checked directly from its definition

$$\sum_b H_{ab} e^{-\frac{\beta E_b}{2}} = 0 \tag{4.37}$$

Therefore, $Q_a^{(E)}$ is an equilibrium state of the system. Consequently,

$$P_a^{(E)} = e^{-\frac{\beta E_a}{2}} Q_a^{(E)} = C e^{-\beta E_a} \tag{4.38}$$

is the corresponding equilibrium distribution for the probabilities of occupancy of the states, and the normalization implies $C = 1/Z$.

Equation (4.27) can be solved formally as

$$Q_a(t) = \sum_b (e^{-Ht})_{ab} Q_b(0) \tag{4.39}$$

where e^{-Ht} is the exponential of the operator H and $Q(0)$ is the initial condition. Using the orthonormal basis of eigenvectors of the symmetric operator H, we may write

$$(e^{-Ht})_{ab} = \sum_{n=0}^{+\infty} e^{-\lambda_n t} Q_a^{(n)} Q_b^{(n)} \tag{4.40}$$

where the $Q^{(n)}$ are the normalized eigenvectors of H

$$\sum_b H_{ab} Q_b^{(n)} = \lambda_n Q_a^{(n)} \tag{4.41}$$

with

$$\sum_{n=0}^{+\infty} Q_a^{(n)} Q_b^{(n)} = \delta_{ab} \tag{4.42}$$

Using the relation (4.26), and isolating the ground state of H, we may write the spectral expansion of the probability as

$$P_a(t) = \frac{e^{-\beta E_a}}{Z} + \sum_{n=1}^{+\infty} e^{-\lambda_n t} c_a^{(n)} \tag{4.43}$$

where the $c_a^{(n)}$ are some real coefficients determined by the initial conditions. When $t \to \infty$, we see that $P_a(t)$ converges to the Boltzmann distribution (4.38). Therefore, a Markovian system with transition rates which satisfy detailed balance always converges to an equilibrium distribution at large time. Note the analogy of Eq. (4.43) with (3.27) in the continuous case.

Equation (4.43) shows that the relaxation time τ_r of the system is given by the inverse of the gap of H

$$\tau_r = \frac{1}{\lambda_1} \tag{4.44}$$

where λ_1 is the smallest non zero eigenvalue, in complete analogy with Eq. (3.22) for the Fokker–Planck case. The whole situation we described above is very similar to the one encountered in section 3.3 on the quantum mechanical representation.

4.3 Kinetic Monte Carlo (KMC)

The master equation is the evolution equation for the probability of the system to be in a given state. It is similar to the continuous space Fokker–Planck equation for the particle distribution. The Fokker–Planck equation can be derived from the Langevin Equation, which is a description of the dynamics in terms of individual trajectories. In the discrete case, there is no equivalent of the Langevin equation, but one can generate trajectories $\{s(t)\}$ in the space of states which exactly sample the master equation by using a technique called *kinetic Monte Carlo* [3]. Assume N copies of the system are evolved stochastically. Let's denote by $N_a(t)$ the number of systems in state a at time t. The probability to find a system in state a is estimated by

$$P_a(t) = \frac{N_a(t)}{N} \tag{4.45}$$

and the master equation (4.8) can be recast in differential form as

$$N_a(t + dt) = N_a(t) + \sum_{b \neq a} k_{ab} N_b(t) dt - \Gamma_a N_a(t) dt \tag{4.46}$$

where

$$\Gamma_a = \sum_{b \neq a} k_{ba} \tag{4.47}$$

is the probability of a particle in state a to jump to any other state b per unit time. This means that $\Gamma_a dt$ is the conditional probability that given the system was in state a at time t, it has jumped away from a at time $t + dt$. Let us define the probability for a particle to have remained in state a during a period t by $R_a(t)$. It is easy to see that

$$R_a(t + dt) = R_a(t) - \Gamma_a dt R_a(t) \tag{4.48}$$

or

$$\frac{dR_a}{dt} = -\Gamma_a R_a \tag{4.49}$$

which can be integrated as

$$R_a(t) = e^{-\Gamma_a t} \tag{4.50}$$

The initial condition is $R_a(0) = 1$ since we assumed that the system was originally in state a. From this equation, it is easy to obtain the *dwell time* probability distribution function. This is the probability distribution of the time spent by the system in a given state a before escaping away from that state for the first time. Indeed, the dwell time probability $P_a^{(D)}(t) dt$ is equal to the probability that given the system was originally in state a, it has remained in state a till time t and then has jumped to a different state at time $t + dt$. It follows that

$$P_a^{(D)}(t)dt = e^{-\Gamma_a t}\Gamma_a dt \tag{4.51}$$

where the factor $\Gamma_a dt$ is the probability of the particle to have escaped from state a during time dt. The mean dwell time is given by

$$\tau_a^{(D)} = \int_0^\infty dt \; t P_a^{(D)}(t) = \frac{1}{\Gamma_a} \tag{4.52}$$

It is the average time the system stays in state a before escaping for the first time to another state. As we shall see, it is directly related to the *first passage time* defined in section 4.1. This important relation between the rate Γ_a of escape from state a and the mean dwell time can be used to compute reaction rates from numerical simulations. As we defined it, $P_a^{(D)}(t)$ is the p.d.f. of the time t the system spends in state a before making a transition to any other state. This allows to define the "trajectory" of a system in the following way. Imagine at time t the system is in state a. The system will jump away from a at time t with probability $P_a^{(D)}(t)$. Once the system makes a transition from a, it jumps to a state b with probability $\frac{k_{ba}}{\Gamma_a}$. This defines the *kinetic Monte Carlo* scheme. The practical implementation of the algorithm goes as follows.

1. Assume the system is in a state $s(t_n) = a$ at time t_n. The system remains in state a during some random time τ after which it makes a transition to some state $s(t_n + \tau) = b$. The time τ is thus the dwell time in a. It is therefore drawn from the distribution $P_a^{(D)}(\tau) = \Gamma_a e^{-\Gamma_a \tau}$. Practically, if we have a random number generator $P_u(r)$ which produces random numbers uniformly distributed between 0 and 1, we can easily generate random numbers distributed with the law $P_a^{(D)}(t)$ using the theorem for the change of variables in probability distributions. According to this theorem, we have

$$P_a^{(D)}(\tau)|d\tau| = P_u(r)|dr| \tag{4.53}$$

so that

$$\left|\frac{dr}{d\tau}\right| = \Gamma_a e^{-\Gamma_a \tau} \tag{4.54}$$

This implies

$$r = e^{-\Gamma_a \tau} \tag{4.55}$$

or equivalently

$$\tau = -\frac{1}{\Gamma_a} \ln r \tag{4.56}$$

So if one draws a random number r uniformly distributed between 0 and 1 and define τ according to Eq. (4.56), then τ is distributed according to the law $P_a^{(D)}(\tau)$.

2. The state $b = s(t_n + 1)$ to which the system will make the transition is chosen with probability $\frac{k_{ba}}{\Gamma_a}$. Practically, a simple way to implement this probability is to align consecutively on $[0, 1]$ all the segments of length $\frac{k_{ba}}{\Gamma_a}$, $b \neq a$, for instance as $\frac{k_{1a}}{\Gamma_a}, \frac{k_{2a}}{\Gamma_a}, \ldots, \frac{k_{Ma}}{\Gamma_a}, \ldots$ They fill up the whole segment $[0, 1]$ since $\Gamma_a = \sum_{b \neq a} k_{ba}$. One then draws uniformly a number r_1 between 0 and 1. This number r_1 falls in between $\sum_{b_1=1}^{b} \frac{k_{b_1 a}}{\Gamma_a}$ and $\sum_{b_1=1}^{b+1} \frac{k_{b_1 a}}{\Gamma_a}$ in $[0, 1]$, $\sum_{b_1=1}^{b} \frac{k_{b_1 a}}{\Gamma_a} < r_1 < \sum_{b_1=1}^{b+1} \frac{k_{b_1 a}}{\Gamma_a}$. The state to which the system makes a transition is the state b indexed by the segment in which r_1 belongs, i.e. satisfying the previous inequality (see Figure 4.1).
3. Go back to 1 and iterate the procedure until final time t is reached.

This kinetic Monte Carlo algorithm generates a trajectory $s(\tau)$ of the system in the space of states $\{a\}$: the system stays in a state for a certain period, then jumps to available states according to the algorithm. It is schematically depicted in Figure 4.2.

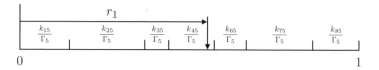

Figure 4.1 Illustration of the choice of the state to which the system jumps. The random number r_1 falls in $\frac{k_{45}}{\Gamma_5}$ so the system jumps to state 4.

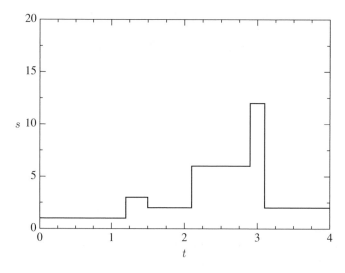

Figure 4.2 A sample MC trajectory in the space of states.

In order to perform ensemble averages with the KMC method, one must generate a large number $\mathcal{N}$ of trajectories $\{s_\alpha(\tau)\}, \alpha \in \{1, \ldots, \mathcal{N}\}$, corresponding to different noise histories, i.e. different samplings of time and state sequences. With this method, all generated trajectories are statistically independent, and therefore, all statistical observables can be computed directly as averages over this ensemble. If for instance A is an observable of the system, which takes the value $A(a)$ when the system is in state a, we have

$$\langle A(t) \rangle = \frac{1}{\mathcal{N}} \sum_{\alpha=1}^{\mathcal{N}} A(s_\alpha(t)) \tag{4.57}$$

where $s_\alpha(t)$ denotes the state of the system at time t for trajectory α.

References

1 (a) Jayachandran, G., Vishal, V., and Pande, V.S. (2006). Using massively parallel simulation and Markovian models to study protein folding: examining the dynamics of the villin headpiece, *J. Chem. Phys.* 124 (16): 164902. (b) Pande, V.S., Beauchamp, K., and Bowman, G.R. (2010). Everything you wanted to know about Markov State Models but were afraid to ask, *Methods* 52 (1): 99.
2 Rogers, L.C.G. and Williams, D. (2000). *Diffusions, Markov Processes, and Martingales, 2e*. Cambridge University Press.
3 (a) Gillespie, D.T. (1976). A general method for numerically simulating the stochastic time evolution of coupled chemical reactions, *J. Comput. Phys.* 22. (b) Gillespie, D.T. (1977). Exact stochastic simulation of coupled chemical reactions, *J. Phys. Chem.* 81 (25): 2340.

5

Path Integrals

Having studied the dynamics of a system from the trajectory (Langevin equation) or from the probability distribution perspective (Fokker–Planck or Schrödinger equation), there is a very powerful method which somehow combines these two points of view, which was first introduced by N. Wiener in probability theory in 1921, then by P.A.M. Dirac in quantum mechanics in 1933 and further developed by R. Feynman in 1948 [1]. This method is called *path integral*. The idea is that the probability for a particle to go from point A to point B in time t is the sum over all paths (physical or not) going from A to B in time t, of the exponential of a certain weight which will be discussed in the following. For simplicity, we will assume throughout this section that the particle follows overdamped Langevin dynamics, with a homogeneous diffusion constant (although everything could be done with a space dependent one). The method we expose is essentially due to Onsager and Machlup [2].

5.1 The Itô Path Integral

Using the Euler-Maruyama discretization scheme (1.91), we have

$$x_{n+1} = x_n + \frac{1}{\gamma}F(x_n, \tau_n)dt + \sqrt{2Ddt}\ \zeta_n \tag{5.1}$$

where the ζ_n are normal Gaussian identically distributed variables $\mathcal{N}(0, 1)$. It follows that the conditional probability for the particle to be at x_{n+1} at time $\tau_{n+1} = \tau_n + dt$, given that it was at x_n at time τ_n, is given by the change of function

$$P(x_{n+1}, \tau_{n+1}|x_n, \tau_n)dx_{n+1} = P(\zeta_n)d\zeta_n \tag{5.2}$$

and since

$$\frac{d\zeta_n}{dx_{n+1}} = \frac{1}{\sqrt{2Ddt}} \tag{5.3}$$

Molecular Kinetics in Condensed Phases: Theory, Simulation, and Analysis,
First Edition. Ron Elber, Dmitrii E. Makarov and Henri Orland.
© 2020 John Wiley & Sons Ltd. Published 2020 by John Wiley & Sons Ltd.

is a constant, it follows that

$$P(x_{n+1}, \tau_{n+1}|x_n, \tau_n) = \frac{1}{\sqrt{2Ddt}} \cdot \frac{1}{\sqrt{2\pi}} e^{-\frac{\xi_n^2}{2}} = \frac{1}{\sqrt{4\pi Ddt}} e^{-\frac{dt}{4D} \left(\frac{x_{n+1}-x_n}{dt} - \frac{1}{\gamma} F(x_n, \tau_n) \right)^2}$$

(5.4)

Therefore, the probability that the particle goes through a given sequence of points $(x_0, \tau_0), (x_1, \tau_1), \ldots, (x_{N-1}, \tau_{N-1}), (x_N, \tau_N)$ is given by

$$P(x_N, \tau_N; x_{N-1}, \tau_{N-1}; \ldots; x_1, \tau_1; x_0, \tau_0) = \left(\frac{1}{4\pi Ddt} \right)^{\frac{N}{2}} e^{-\frac{dt}{4D} \sum_{n=0}^{N-1} \left(\frac{x_{n+1}-x_n}{dt} - \frac{1}{\gamma} F(x_n, \tau_n) \right)^2}$$

(5.5)

If we are now interested in the probability for the particle to start at x_i at time 0 and to end at x_f at time t_f, we need to integrate over all intermediate points $x_1, \ldots, x_{N-1}$, and we have thus

$$P(x_f, t_f|x_i, 0) = \frac{1}{C} \int dx_1 \ldots dx_{N-1} e^{-\frac{dt}{4D} \sum_{n=0}^{N-1} \left(\frac{x_{n+1}-x_n}{dt} - \frac{1}{\gamma} F(x_n, \tau_n) \right)^2}$$

(5.6)

where x_0 is constrained to be equal to x_i and x_N to x_f, and $C = (4\pi Ddt)^{\frac{N}{2}}$ is a constant. The collection of points $\{(x_i, 0), (x_1, \tau_1), \ldots, (x_{N-1}, \tau_{N-1}), (x_f, t_f)\}$ can be viewed as a discretized path joining x_i to x_f. Equation (5.6) lends itself naturally to the continuous limit. Indeed, when $dt \to 0$, the succession of points makes up a continuous trajectory, and we can replace the discrete sum by an integral and the above equation becomes

$$P(x_f, t_f|x_i, 0) = \int_{x(0)=x_i}^{x(t_f)=x_f} \mathcal{D}x(\tau) \, e^{-\frac{1}{4D} \int_0^{t_f} d\tau \left(\frac{dx}{d\tau} - D\beta F(x(\tau), \tau) \right)^2}$$

(5.7)

where we use the notations

$$\frac{dx}{dt} = \lim_{dt \to 0} \frac{x_{n+1} - x_n}{\tau_{n+1} - \tau_n}$$

(5.8)

and

$$\mathcal{D}x(\tau) = \lim_{dt \to 0} \frac{1}{C} \prod_{n=1}^{N-1} dx_n$$

(5.9)

Note that despite the use of the derivative notation above, it does not mean that the paths are differentiable. Indeed, as we saw in section 1.5, the increment dx is of order $\sqrt{dt}$ for the relevant paths, and thus the derivative dx/dt does not exist in the limit $dt \to 0$.

The probability for the particle to go from x_i at time 0 to x_f at time t_f is the sum, over all continuous paths connecting x_i and x_f, of the exponential of a functional of the path, called the *Onsager-Machlup action*, given by

$$S = \frac{1}{4D} \int_0^{t_f} d\tau \left(\frac{dx}{d\tau} - D\beta F(x(\tau), \tau) \right)^2 \tag{5.10}$$

Equation (5.7) defines a *path integral* or *functional integral*. Equation (5.6) is the discretized Onsager-Machlup form of the path integral. The integral in (5.10) is to be understood as an Itô integral (1.94). Consequently, expanding S in Eq. (5.10), the integral of the cross term

$$I_I = \int_0^{t_f} d\tau \frac{dx}{d\tau} F(x(\tau), \tau) \tag{5.11}$$

would have had a different expression if we had used the Stratonovich discretization of the Langevin equation. We will detail that below.

In the case of a conservative force, there is a further simplification of this equation. Indeed, for $F = -\frac{\partial U}{\partial x}$ we have

$$\begin{aligned} S &= \frac{1}{4D} \int_0^{t_f} d\tau \left(\frac{dx}{d\tau} + D\beta \frac{\partial U}{\partial x} \right)^2 \\ &= \frac{1}{4D} \int_0^{t_f} d\tau \left(\left(\frac{dx}{dt} \right)^2 + \left(D\beta \frac{\partial U}{\partial x} \right)^2 \right) + \frac{\beta}{2} \int_0^{t_f} d\tau \frac{dx}{d\tau} \frac{\partial U}{\partial x} \end{aligned} \tag{5.12}$$

Using the Itô formula (1.94) for the last integral above,

$$\int_0^{t_f} d\tau \frac{dx}{d\tau} \frac{\partial U}{\partial x} = U(x_f) - U(x_i) - D \int_0^{t_f} dt \frac{\partial^2 U}{\partial x^2} \tag{5.13}$$

we obtain

$$P(x_f, t_f | x_i, 0) = e^{-\frac{\beta}{2}(U(x_f) - U(x_i))} \int_{x(0) = x_i}^{x(t_f) = x_f} \mathcal{D}x(\tau) \, e^{-\int_0^{t_f} d\tau \left(\frac{1}{4D} \left(\frac{dx}{d\tau} \right)^2 + DV(x(\tau)) \right)} \tag{5.14}$$

where $V(x)$ is the effective potential defined in Eq. (3.5)

$$V(x) = \left(\frac{\beta}{2} \frac{\partial U}{\partial x} \right)^2 - \frac{\beta}{2} \frac{\partial^2 U}{\partial x^2} \tag{5.15}$$

The second derivative in the potential above comes from the Itô formula, and is a correction to the standard chain rule.

Connecting back to quantum mechanics, we see that the path integral in (5.14) is the Feynman path integral representation for the evolution operator of the quantum Hamiltonian defined in Eq. (3.7)

$$\int_{x(0) = x_i}^{x(t_f) = x_f} \mathcal{D}x(\tau) \, e^{-\int_0^{t_f} d\tau \left(\frac{1}{4D} \left(\frac{dx}{d\tau} \right)^2 + DV(x(\tau)) \right)} = \langle x_f | e^{-Ht_f} | x_i \rangle \tag{5.16}$$

so that

$$P(x_f, t_f | x_i, 0) = e^{-\frac{\beta}{2}(U(x_f) - U(x_i))} \langle x_f | e^{-H t_f} | x_i \rangle \tag{5.17}$$

which is identical to the fundamental equation (3.13). This last equation allows to easily prove detailed balance. Indeed, the Hamiltonian H being real and symmetric (Hermitian), we have

$$\langle x_f | e^{-H t_f} | x_i \rangle = \langle x_i | e^{-H t_f} | x_f \rangle \tag{5.18}$$

It follows that

$$\frac{P(x_f, t_f | x_i, 0)}{P(x_i, t_f | x_f, 0)} = e^{-\beta(U(x_f) - U(x_i))} \tag{5.19}$$

which proves detailed balance for any initial and final conditions.

5.2 The Stratonovich Path Integral

It is instructive to show how one can recover all these expressions using the Stratonovich discretization. Indeed, in that case, the discretized equation becomes

$$x_{n+1} = x_n + \frac{1}{2\gamma}(F(x_{n+1}, \tau_{n+1}) + F(x_n, \tau_n))dt + \sqrt{2Ddt}\, \zeta_n \tag{5.20}$$

and the change of function Eq. (5.3) imply

$$\begin{aligned}
\frac{d\zeta_n}{dx_{n+1}} &= \frac{1}{\sqrt{2Ddt}}\left(1 - \frac{1}{2\gamma}\frac{\partial F(x_{n+1}, \tau_{n+1})}{\partial x_{n+1}}dt\right) \\
&\approx \frac{1}{\sqrt{2Ddt}}e^{-\frac{1}{2\gamma}\frac{\partial F(x_{n+1}, \tau_{n+1})}{\partial x_{n+1}}dt}
\end{aligned} \tag{5.21}$$

in the limit $dt \to 0$ and therefore the discretized path integral (5.6) becomes

$$P(x_f, t_f | x_i, 0)$$
$$= \frac{1}{C}\int dx_1 \dots dx_{N-1} e^{-dt\sum_{n=0}^{N-1}\left(\frac{1}{4D}\left(\frac{x_{n+1}-x_n}{dt} - \frac{D\beta}{2}(F(x_{n+1},\tau_{n+1})+F(x_n,\tau_n))\right)^2 + \frac{D\beta}{2}\frac{\partial F(x_{n+1},\tau_{n+1})}{\partial x_{n+1}}\right)} \tag{5.22}$$

which in the continuous limit yields

$$P(x_f, t_f | x_i, 0) = \int_{x(0)=x_i}^{x(t_f)=x_f} \mathscr{D}x(\tau)\, e^{-\int_0^{t_f}\left(\frac{1}{4D}\left(\frac{dx}{d\tau} - D\beta F(x(\tau),\tau)\right)^2 + \frac{D\beta}{2}\frac{\partial F(x(\tau),\tau)}{\partial x(\tau)}\right)\bullet d\tau} \tag{5.23}$$

where now the exponent involves Stratonovich integrals. If we consider the following two terms of the exponent above

$$J_S = \frac{\beta}{2} \left(\int_0^{t_f} \frac{dx}{d\tau} F(x(\tau), \tau) \bullet d\tau - D \int_0^{t_f} \frac{\partial F(x(\tau), \tau)}{\partial x(\tau)} \bullet d\tau \right) \tag{5.24}$$

we see, using Eq. (1.110) that

$$J_S = \frac{\beta}{2} I_I \tag{5.25}$$

where I_I is the Itô integral

$$I_I = \int_0^{t_f} d\tau \frac{dx}{d\tau} F(x(\tau), \tau) \tag{5.26}$$

This last equation shows the exact equivalence of the Itô (5.7) and Stratonovich (5.23) path integral representations of the transition probabilities.

To be complete, let us mention the case of underdamped Langevin dynamics. In that case, using the same formalism as above, the probability for the particle to be at point x_f with momentum p_f at time t_f, given that it was at x_i with momentum p_i at time 0, can be written as an Itô path integral as

$$P(x_f, p_f, t_f | x_i, p_i, 0) = \int \mathscr{D} x(\tau) e^{-\frac{1}{4\gamma k_B T} \left(m \frac{d^2 x}{dt^2} + \gamma \frac{dx}{dt} + F(x(\tau), \tau) \right)^2} \tag{5.27}$$

with the boundary conditions on the paths

$$x(0) = x_i$$
$$\frac{dx(0)}{dt} = \frac{p_i}{m}$$
$$x(t_f) = x_f$$
$$\frac{dx(t_f)}{dt} = \frac{p_f}{m} \tag{5.28}$$

These path integral representations of the dynamics of the system allow for powerful computational techniques, ranging from statistically exact sampling (Monte Carlo sampling, Transition Path sampling, etc.) to approximate techniques, such as perturbation theory or dominant path theory (semi-classical methods). This will be discussed in chapter 7.

References

1 (a) Feynman, R.P. and Hibbs, A.R. (1965). *Quantum Mechanics and Path Integrals*. McGraw-Hill. (b) Negele, J.W. and Orland, H. (1998). *Quantum Many-Particle Systems (Advanced Books Classics)*. Perseus Books.
2 Onsager, L. and Machlup, S. (1953). Fluctuations and irreversible processes, *Phys. Rev.* 91 (6): 1505.

6

Barrier Crossing

Most physical systems can be viewed, in a schematic way, as undergoing either diffusion (possibly in an inhomogeneous medium, with space dependent diffusion constant), or barrier crossing, i.e. going from one metastable state to another metastable state through a free energy barrier. Diffusion phenomena are characterized by scaling exponents ν, which specify the typical size R of the region visited by the particle during time t as

$$R^2 \sim Dt^\alpha \tag{6.1}$$

where D is the diffusion coefficient, and α is the diffusion exponent. In normal diffusion phenomena, $\alpha = 1$. If the exponent is not equal to 1, $\alpha \neq 1$, the phenomenon is called *anomalous diffusion*. It is called super-diffusion if $\alpha > 1$ and sub-diffusion if $\alpha < 1$. Both phenomena are observed experimentally, in polymeric or colloid fluids, in active cellular transport, in hydrodynamics and in many other physical, chemical or biological phenomena.

Since the topic of this book is mostly chemical and biochemichal reactions, we will study in more details the case of *barrier crossing*. This subject will be further studied in chapter 10. In this case, the system is prepared in a given state A, usually metastable, i.e. a local minimum of the free energy, and makes a spontaneous thermally driven transition to another metastable state B, by going over a free energy barrier. This is the paradigm for many chemical reactions, which display an activation barrier, and also for many biochemical reactions, such as protein folding, allosteric transitions, etc. In the case of high barrier (compared to the thermal energy $k_B T$), the system is characterized by two very separate time scales, which we will study in more details in sections 6.2 and 6.3.

6.1   First Passage Time and Transition Rate

In order to study barrier crossing and transition rates, it is useful to introduce the notion of *first passage time*. This matter will be further revisited in Chapter 13 and subsequent ones. For a thorough review on the subject, we refer the

Molecular Kinetics in Condensed Phases: Theory, Simulation, and Analysis,
First Edition. Ron Elber, Dmitrii E. Makarov and Henri Orland.
© 2020 John Wiley & Sons Ltd. Published 2020 by John Wiley & Sons Ltd.

reader to the book by S. Redner [1]. As its name indicates, the first passage time is the first time the particle crosses a given boundary. The first passage time is closely related to the mean dwell time defined in 4.3. As it was defined, the dwell time is the time the system spends in a given state A before leaving A for the first time to any other state. The first passage time from A to B is the time the system stays in A before reaching B for the first time. In section 4.3, we showed in Eq. (4.52) that the mean dwell time in A is the inverse of the rate of escape Γ_A from A. Similarly, it can be easily shown that the mean first passage time τ_1 from A to B is equal to the inverse of the transition rate k_{BA} from A to B

$$\tau_1 = \frac{1}{k_{BA}} \tag{6.2}$$

In the case of a particle subject to Langevin dynamics, let us define a certain domain Ω in space and denote its boundary as $S = \partial\Omega$. The particle is supposed to start at some point x_i inside Ω at $t = t_0$. We may define the first passage time of the particle at S as the first time it crosses S. The probability that the particle is still in Ω at time $t = t_0 + \tau$, also called *survival probability*, is given by

$$P_\Omega(\tau, x_i) = \int_\Omega dx P_a(x, t | x_i, t_0) \tag{6.3}$$

where *absorbing boundary conditions* are imposed on the boundary S, that is $P_a(x, t | x_i, t_0) = 0$ if x is in S. In other words, as soon as the particle touches S, it is absorbed, thus guaranteeing that only trajectories of particles that have not left the domain Ω and come back to it contribute to P_a. Let us emphasize that $P_a(x, t | x_i, t_0)$ satisfies the standard Fokker–Planck (or adjoint Fokker–Planck) equation, the absorbing condition being just a boundary condition imposed on the Fokker–Planck equation.

With this boundary condition, as long as the particle is in Ω, it is alive and has not crossed S yet, so $P_\Omega(\tau, x_i)$ is the probability that the first passage time is larger than τ

$$P_\Omega(\tau, x_i) = \text{Probability}(1^{st} \text{ passage time} > \tau)$$
$$= \langle \theta(1^{st} \text{ passage time} - \tau) \rangle \tag{6.4}$$

where the bracket denotes an average over all paths starting at x_i at time t_0, satisfying the absorbing boundary conditions on $S = \partial\Omega$. The function $\theta(u)$ is the Heavyside function, equal to 1 if its argument $u \geq 0$ and to 0 otherwise.

The first passage time $\tau(x_i)$ is a random variable (which depends on the initial point x_i), and denoting its probability distribution by $P_1(\tau, x_i)$, from Eq. (6.4) we have

$$P_1(\tau, x_i) = -\frac{\partial P_\Omega(\tau, x_i)}{\partial \tau}$$
$$= -\int_\Omega dx \frac{\partial P_a(x, t_0 + \tau | x_i, t_0)}{\partial \tau} \tag{6.5}$$

obtained by taking the derivative of (6.4) and using the identity

$$\frac{d\theta(t)}{dt} = \delta(t) \tag{6.6}$$

6.1.1 Average Mean First Passage Time

An important quantity is the *mean first passage time* $\tau_1(x_i)$ defined as

$$
\begin{aligned}
\tau_1(x_i) &= \int_0^\infty d\tau \, \tau P_1(\tau, x_i) \\
&= -\int_0^\infty d\tau \, \tau \frac{\partial P_\Omega(\tau, x_i)}{\partial \tau} \\
&= \int_0^\infty d\tau \, P_\Omega(\tau, x_i) \\
&= \int_{t_0}^\infty dt \int dx P_a(x, t | x_i, t_0)
\end{aligned}
$$

where $t = t_0 + \tau$. In the derivation above, we have performed an integration by part and used the fact that $P_\Omega(\tau = \infty, x_i) = 0$. The vanishing of this probability is a direct result of Eq. (6.4) if one assumes that the first passage time is finite.

We can now easily obtain a partial differential equation for the mean first passage time. Indeed, we have

$$D\frac{\partial^2 \tau_1(x_i)}{\partial x_i^2} = D\int_{t_0}^\infty dt \int dx \frac{\partial^2 P_a(x, t | x_i, t_0)}{\partial x_i^2}$$

and since this is a derivative with respect to the initial condition, we may use the adjoint Fokker–Planck equation (2.32)

$$D\frac{\partial^2 \tau_1(x_i)}{\partial x_i^2} = \int_{t_0}^\infty dt \int dx \left(-\frac{dP_a(x, t | x_i, t_0)}{dt_0} - D\beta F(x_i)\frac{\partial P_a(x, t | x_i, t_0)}{\partial x_i} \right) \tag{6.7}$$

Since the force $F(x)$ is independent of time, the probability $P_a(x, t | x_i, t_0)$ is a function of $t - t_0$ and thus $\frac{dP_a(x, t | x_i, t_0)}{dt_0} = -\frac{dP_a(x, t | x_i, t_0)}{dt}$. Inserting this identity in (6.7) and using the relations

$$\int_{t_0}^\infty dt \int_\Omega dx \frac{dP_a(x, t | x_i, t_0)}{dt} = \int_\Omega dx (P_a(x, \infty | x_i, t_0) - P_a(x, t_0 | x_i, t_0)) \tag{6.8}$$

and

$$\int_\Omega dx P_a(x, \infty | x_i, t_0) = P_\Omega(\tau = \infty, x_i) = 0 \tag{6.9}$$

$$\int_\Omega dx P_a(x, t_0 | x_i, t_0) = 1 \tag{6.10}$$

we obtain

$$D\left(\frac{\partial^2 \tau_1(x_i)}{\partial x_i^2} + \beta F(x_i)\frac{\partial \tau_1(x_i)}{\partial x_i}\right) = -1 \qquad (6.11)$$

Equation (6.9) is a consequence of Eq. (6.4), which implies that the probability of survival of the particle in Ω vanishes for infinite time, whereas Eq. (6.10) follows from the initial condition $P_a(x, t_0|x_i, t_0) = \delta(x - x_i)$. This partial differential equation is to be supplemented with the boundary condition

$$\tau_1(x_i) = 0 \text{ for any } x_i \text{ in the boundary } S \qquad (6.12)$$

In the case of a conservative force, $F = -\frac{\partial U}{\partial x}$, multiplying Eq. (6.11) by $e^{-\beta U}$ we have

$$D\left(e^{-\beta U(x)}\frac{\partial^2 \tau_1(x)}{\partial x^2} - \beta e^{-\beta U(x)}\frac{\partial U}{\partial x}\frac{\partial \tau_1(x)}{\partial x}\right) = -e^{-\beta U(x)} \qquad (6.13)$$

or

$$\frac{\partial}{\partial x}\left(e^{-\beta U}\frac{\partial \tau_1(x)}{\partial x}\right) = -\frac{1}{D}e^{-\beta U} \qquad (6.14)$$

In one dimension, this equation can be further simplified by integration. Indeed, in one dimension, if the domain Ω is defined by the segment $[a, b]$, its boundary consists of the two points $S = \{a, b\}$ and we can integrate the above equation to obtain

$$\frac{\partial \tau_1(x)}{\partial x} = -\frac{1}{D}e^{\beta U(x)}\int_a^x dz\, e^{-\beta U(z)} + A_1 e^{\beta U(x)} \qquad (6.15)$$

where A_1 is a constant of integration. One can integrate again the above equation and obtain

$$\tau_1(x) = -\frac{1}{D}\int_a^x dy\, e^{\beta U(y)}\int_a^y dz\, e^{-\beta U(z)} + A_1\int_a^x dy\, e^{\beta U(y)} + A_2 \qquad (6.16)$$

where A_2 is a second integration constant. The two integration constants A_1 and A_2 are determined by requiring that $\tau(a) = \tau(b) = 0$. This implies

$$A_2 = 0$$

$$A_1 = \frac{1}{D}\frac{\int_a^b dy\, e^{\beta U(y)}\int_a^y dz\, e^{-\beta U(z)}}{\int_a^b dy\, e^{\beta U(y)}} \qquad (6.17)$$

6.1.2 Distribution of First Passage Time

Similarly, the distribution of first passage time satisfies a partial differential equation. Indeed, from Eq. (6.5), we have

$$
\begin{aligned}
P_1(\tau, x_i) &= -\int_\Omega dx \, \frac{\partial P_a(x, t | x_i, t_0)}{\partial t} & (6.18) \\
&= \int_\Omega dx \, \frac{\partial P_a(x, t | x_i, t_0)}{\partial t_0} \\
&= -D \int_\Omega dx \left(\frac{\partial^2 P_a(x, t | x_i, t_0)}{\partial x_i^2} + \beta F(x_i) \frac{\partial P_a(x, t | x_i, t_0)}{\partial x_i} \right) \\
&= -D \left(\frac{\partial^2 P_\Omega(\tau, x_i)}{\partial x_i^2} + \beta F(x_i) \frac{\partial P_\Omega(\tau, x_i)}{\partial x_i} \right)
\end{aligned}
$$

where $t = t_0 + \tau$, and using Eq. (6.5) we get the equation for p.d.f. of the first passage time τ

$$
\frac{\partial P_1(\tau, x)}{\partial \tau} = D \left(\frac{\partial^2 P_1(\tau, x)}{\partial x^2} + \beta F(x) \frac{\partial P_1(\tau, x)}{\partial x} \right) \tag{6.19}
$$

The initial condition on P_1 is a bit more involved. From Eq. (6.18), using the Fokker–Planck equation, we have

$$
P_1(\tau, x_i) = -D \int_\Omega dx \, \frac{\partial}{\partial x} \left(\frac{\partial P_a}{\partial x} - \beta F(x) P_a(x, t_0 + \tau | x_i, t_0) \right) \tag{6.20}
$$

Using Stokes theorem, the integral over Ω can be reduced to an integral over the boundary $S = \partial \Omega$

$$
P_1(\tau, x_i) = -D \int_S d\vec{\sigma}_x \left(\frac{\partial P_a}{\partial x} - \beta F(x) P_a(x, t_0 + \tau | x_i, t_0) \right) \tag{6.21}
$$

where $d\vec{\sigma}_x$ is the surface element on S (oriented to the exterior of Ω). Since $P_a(x, t | x_i, t_0)$ vanishes on the boundary S, the second term on the r.h.s above is zero and we have

$$
P_1(\tau, x_i) = -D \int_S d\vec{\sigma}_x \, \frac{\partial P_a(x, t_0 + \tau | x_i, t_0)}{\partial x} \tag{6.22}
$$

For $\tau = 0$, if x_i is not on S, the absorbing boundary condition is irrelevant and

$$
P_a(x, t_0 | x_i, t_0) = \delta(x - x_i) \tag{6.23}
$$

Therefore, the initial condition on P_1 is

$$
P_1(0, x) = -D \int_S d\vec{\sigma}_y \, \frac{\partial}{\partial y} \delta(y - x) \tag{6.24}
$$

In d-dimensional space, the partial derivative is a gradient and the above formula should be written as

$$P_1(0, \vec{x}) = -D \int_S d\vec{\sigma}_y \nabla_y \delta(\vec{y} - \vec{x}) \tag{6.25}$$

For $\tau \neq 0$, the p.d.f. must satisfy the boundary condition

$$P_1(\tau, x) = 0 \text{ for any } x \text{ in } S \tag{6.26}$$

which expresses the fact that if the starting point x belongs to the boundary S, the first passage time τ to S is equal to 0.

6.1.3 The Free Particle Case

Let us illustrate these concepts with a free particle. Assume the particle is in the $x \geq 0$ half-plane, and we ask what is the first passage time at $x = 0$. The boundary is thus reduced to the point $S = \{0\}$. The p.d.f. for the particle with absorbing boundary condition at $x = 0$ can be obtained by the method of images [2]. Indeed, in order to solve the free Fokker–Planck equation

$$\frac{\partial P_a}{\partial t} = D \frac{\partial^2 P_a}{\partial x^2} \tag{6.27}$$

with the initial condition $P_a(x, t_0 | x_i, t_0) = \delta(x - x_i)$ and the absorbing boundary condition $P_a(0, t | x_i, t_0) = 0$, we can use the linearity of the FP equation. If we consider the solution of the FP equation with initial condition at $-x_i$, symmetric of the initial point x_i with respect to the boundary point 0, the difference of the two solutions

$$P_a(x, t | x_i, t_0) = \frac{1}{\sqrt{4\pi D(t - t_0)}} \left(e^{-\frac{(x-x_i)^2}{4D(t-t_0)}} - e^{-\frac{(x+x_i)^2}{4D(t-t_0)}} \right) \tag{6.28}$$

is also a solution of the FP equation which vanishes at $x = 0$ and thus satisfies the correct absorbing boundary condition. Note that the initial condition is

$$P_a(x, t_0 | x_i, t_0) = \delta(x - x_i) - \delta(x + x_i)$$

and since $x > 0$, the second δ–function is irrelevant as it vanishes identically. From Eq. (6.3), the survival probability in the positive half-plane is given by

$$P_\Omega(\tau, x_i) = \int_0^{+\infty} \frac{dx}{\sqrt{4\pi D(t - t_0)}} \left(e^{-\frac{(x-x_i)^2}{4D(t-t_0)}} - e^{-\frac{(x+x_i)^2}{4D(t-t_0)}} \right) \tag{6.29}$$

Performing the change of variable $u = \frac{x \mp x_i}{\sqrt{4D(t-t_0)}}$ respectively in each integral, we obtain

$$P_\Omega(\tau, x_i) = \text{Erf} \left(\frac{x_i}{\sqrt{4D\tau}} \right) \tag{6.30}$$

where Erf is the *error function* defined by

$$\text{Erf}(u) = \frac{2}{\sqrt{\pi}} \int_0^u dx \, e^{-x^2} \tag{6.31}$$

Taking a derivative of (6.30) with respect to τ, and using the identity

$$\frac{d}{du} \text{Erf}(u) = \frac{2}{\sqrt{\pi}} e^{-u^2} \tag{6.32}$$

the first-passage time p.d.f is given by

$$P_1(\tau, x) = -\frac{\partial P_\Omega(\tau, x)}{\partial \tau}$$

$$= \frac{x}{\sqrt{4\pi D \tau^3}} e^{-\frac{x^2}{4D\tau}} \tag{6.33}$$

For any $x \neq 0$, this p.d.f. has an essential singularity at $\tau = 0$, and decays as $\tau^{-3/2}$ at large time. In addition, it is properly normalized as can be seen from the integral

$$\int_0^{+\infty} d\tau P_1(\tau, x) = \int_0^{+\infty} d\tau \frac{x}{\sqrt{4\pi D \tau^3}} e^{-\frac{x^2}{4D\tau}} \tag{6.34}$$

and performing the change of variable $z = \frac{x^2}{4D\tau}$ we have

$$\int_0^{+\infty} d\tau P_1(\tau, x) = \int_0^{+\infty} \frac{dz}{\sqrt{\pi z}} e^{-z}$$

$$= 1$$

The result (6.33) for P_1 could have easily been obtained from Eqs. (6.19) and (6.24). Indeed, in the free case we have

$$P_1(\tau, x) = \int_0^{+\infty} \frac{dy}{\sqrt{4\pi D \tau}} e^{-\frac{(x-y)^2}{4D\tau}} P_1(0, y) \tag{6.35}$$

and Eq. (6.24) becomes

$$P_1(0, y) = -D\delta'(y) \tag{6.36}$$

which when inserted in (6.35) yields back Eq. (6.33). Note that the expression (6.33) satisfies the boundary condition (6.26) since it vanishes at $x = 0$ for $\tau \neq 0$.

6.1.4 Conservative Force

In the case of a conservative force, we can use the same technique as before and obtain

$$\frac{\partial P_1(\tau, x)}{\partial \tau} = D e^{\beta U(x)} \frac{\partial}{\partial x} \left(e^{-\beta U(x)} \frac{\partial P_1(\tau, x)}{\partial x} \right) \tag{6.37}$$

Since P_1 is the p.d.f. of the first passage time, it is normalized as

$$\int_0^\infty d\tau \, P_1(\tau, x) = 1 \tag{6.38}$$

Using the transformation

$$P_1(t, x) = e^{\frac{\beta U(x)}{2}} \Psi(x, t) \tag{6.39}$$

Eq. (6.19) takes the form of the imaginary time Schrödinger equation of Eq. (3.14)

$$\frac{d\Psi}{dt} = D \left(\frac{\partial^2 \Psi}{\partial x^2} - V(x)\Psi \right) \tag{6.40}$$

with

$$V(x) = \left(\frac{\beta}{2} \frac{\partial U}{\partial x} \right)^2 - \frac{\beta}{2} \frac{\partial^2 U}{\partial x^2} \tag{6.41}$$

and therefore, one may use the same spectral representation as for the probability distribution $P(x, t)$ and obtain

$$P_1(t, x) = e^{\frac{\beta U(x)}{2}} \sum_{n=0}^\infty e^{-E_n t} c_n \Psi_n(x) \tag{6.42}$$

where E_n and $\Psi_n(x)$ are the eigenvalues and eigenstates of H as defined in Eq. (3.14) and the coefficients c_n are determined by the initial conditions. As was shown in section 3.3, the ground state energy and wave function are given by

$$E_0 = 0$$

$$\Psi_0(x) = \frac{e^{-\frac{\beta U(x)}{2}}}{\sqrt{Z}} \tag{6.43}$$

so that the expansion (6.42) becomes

$$P_1(t, x) = \frac{c_0}{\sqrt{Z}} + e^{\frac{\beta U(x)}{2}} \sum_{n=1}^\infty e^{-E_n t} c_n \Psi_n(x)$$

$$\sim_{t \to \infty} \frac{c_0}{\sqrt{Z}} + c_1 e^{-E_1 t} e^{\frac{\beta U(x)}{2}} \Psi_1(x) + \dots \tag{6.44}$$

Due to the normalization condition (6.38), the constant c_0 is necessarily 0, and

$$P_1(t, x) \sim e^{\frac{\beta U(x)}{2}} \sum_{n=1}^\infty e^{-E_n t} c_n \Psi_n(x) \tag{6.45}$$

with the initial condition (6.24). In addition, the coefficients c_n must be chosen so that the boundary condition (6.26) is satisfied at any time.

6.2 Kramers Transition Time: Average and Distribution

Chemical reactions can often be viewed as a transition between two states A and B of a system. For instance it can be the simple isomerization of a molecule as shown on Figure 6.1, or a more complex bimolecular chemical reaction, or the transition of a protein from the folded to the unfolded state F $\leftrightarrows$ U. If one uses reaction coordinates, simple reactions can often be described as a transition in a double well type of landscape (see Figure 1.3). In that case, at thermodynamic equilibrium, the system makes random transitions from A to B, with rate constant k_{BA} and random transitions from B to A, with rate constant k_{AB}. As can be seen in Figure 1.2, if the barrier height is large compared to $k_B T$, the system spends most of its time in one of the two states, and makes rapid transitions between them. Indeed, it appears that for high barrier, there are two well separated time scales in the system. One time scale denoted by τ_K is the long time scale which characterizes the time spent by the system in each of the metastable states A and B. The other one denoted by τ_C, is the short time scale which characterizes the time it takes for the system to make a transition from A to B. As can be seen in Figure 1.2, and as we will show in the following, the *crossing time* τ_C is much shorter than τ_K

$$\tau_C \ll \tau_K \tag{6.46}$$

The physical picture which emerges is that of the system being trapped in one of the states A or B where it samples the bottom of the potential well for a long typical time τ_K before making a transition to the other well in a fast jump of typical time τ_C.

The formalism presented in the previous section allows the computation of these time scales, and even their distribution. The first such calculation is due to Kramers [2, 2], who derived the famous Kramers formula for the transition rate for barrier crossing. This is a simplified one-dimensional model of barrier crossing. The one-dimensional degree of freedom should be viewed as the reaction coordinate and the energy $U(x)$ is to be interpreted as the free energy or potential of mean force as a function of the reaction coordinate. We present here the calculation of the escape rate from a potential well for the Langevin equation. The escape rate in the case of the Generalized Langevin Equation with a memory kernel will be studied in chapter 10. We assume that there is a unique metastable state A, and that having passed the barrier, the system escapes to infinity. The shape of the potential is typically that of Figure 6.1, i.e. a potential well around point A, a barrier with coordinate $x^{\ddagger}$ at point Z, and a potential decreasing to $-\infty$ on the right, beyond C.

We assume that there is a population of thermalized particles in the left well A. This notion is to be taken with a grain of salt, as there is no thermodynamic equilibrium of the system localized in the well A, since the particles can thermally go over the barrier Z and leak out to the right of Z. So if the barrier $U^{\ddagger}$

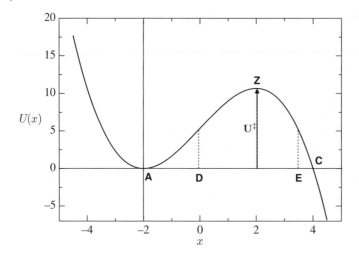

Figure 6.1 Kramers potential.

is larger than the thermal energy $U^{\ddagger} \gg k_B T$, there is a weak current of particles from A to the C region: the system is in a quasi-stationnary state. In 1940, Kramers calculated the rate at which particles escape from A to the C region. We first show a derivation close to the original one, then show how it can be generalized using the mean first passage time described above. In both cases, we study the limit of an overdamped system with a high barrier. We denote by $x_A, x^{\ddagger}$ and x_c the coordinates of points A, Z and C.

6.2.1 Kramers Derivation

The first assumption is that the system is almost at equilibrium around A, i.e. the leaking current q of particles flowing from A to C is small. The Fokker–Planck equation (2.19) reads

$$\frac{dP}{dt} = D\frac{\partial}{\partial x}\left(e^{-\beta U(x)}\frac{\partial}{\partial x}(e^{\beta U(x)}P)\right) \tag{6.47}$$

and the current is given by

$$q = -De^{-\beta U(x)}\frac{\partial}{\partial x}(e^{\beta U(x)}P) \tag{6.48}$$

which implies

$$\frac{\partial}{\partial x}(e^{\beta U(x)}P) = -\frac{q}{D}e^{\beta U(x)} \tag{6.49}$$

Since the system is quasi-stationnary, $\frac{dP}{dt}$ is essentially 0, which implies that the current is time and space independent. We can then integrate Eq. (6.49) between A and C

$$e^{\beta U(x)} P \Big|_{x_A}^{x_C} = -\frac{q}{D} \int_{x_A}^{x_C} dx \; e^{\beta U(x)} \tag{6.50}$$

Since the system is at quasi-equilibrium around A, the probability to find the particle in the C region is very small, and thus we may neglect $P(x_C)$ in the equation above, i.e. $P(x_C) \approx 0$. The current is thus given by

$$q = D \frac{e^{\beta U(x_A)} P(x_A)}{\int_{x_A}^{x_C} dx \; e^{\beta U(x)}} \tag{6.51}$$

On the other hand, q is the probability current. It is the flow of probability per unit time and per unit length. It is thus equal to the probability of the system to be in the quasi-equilibrium state A multiplied by the escape rate from A. We have thus

$$q = P_A k_A \tag{6.52}$$

where P_A is the probability for the particle to be in the basin of A and k_A is the escape rate from A. The probability P_A is computed as the integral of the Boltzmann weight around A, that is in some interval of size Δ, which is the typical size of the basin of attraction of A

$$P_A = \frac{1}{Z_A} \int_{x_A - \Delta}^{x_A + \Delta} dx \; e^{-\beta U(x)} \tag{6.53}$$

In the above equation, $U(x)$ is minimal around x_A and one may thus expand it to second order as

$$U(x) \approx U(x_A) + \frac{1}{2} U''(x_A)(x - x_A)^2 \tag{6.54}$$

where $U''(x_A) > 0$ and there is no first order term in (6.54) since x_A is a minimum. Replacing the expansion in (6.53) we obtain

$$P_A = \frac{e^{-\beta U(x_A)}}{Z_A} \int_{x_A - \Delta}^{x_A + \Delta} dx \; e^{-\frac{\beta U''(x_A)}{2}(x - x_A)^2}$$

$$= P(x_A) \left(\frac{2\pi}{\beta U''(x_A)} \right)^{\frac{1}{2}} \tag{6.55}$$

Similarly, in Eq. (6.51), we have to compute the integral $\int_{x_A}^{x_C} dx \; e^{\beta U(x)}$. In this domain between A and C, the function $U(x)$ is maximal around the top of the barrier Z. We may thus expand it to second order around $x^{\ddagger}$ as

$$U(x) \approx U(x^{\ddagger}) + \frac{1}{2} U''(x^{\ddagger})(x - x^{\ddagger})^2 \tag{6.56}$$

where $U''(x^{\ddagger}) < 0$ and there is no first order term in (6.54) since $x^{\ddagger}$ is a maximum. The integral can be performed and we obtain

$$\int_{x_A}^{x_C} dx\, e^{\beta U(x)} \approx e^{\beta U(x^{\ddagger})} \int_{x_A}^{x_C} dx\, e^{-\frac{\beta |U''(x^{\ddagger})|}{2}(x-x^{\ddagger})^2}$$

$$= e^{\beta U(x^{\ddagger})} \left(\frac{2\pi}{\beta |U''(\ddagger)|} \right)^{\frac{1}{2}} \tag{6.57}$$

Inserting these two equalities in Eq. (6.52), we obtain

$$k_A = \frac{q}{P_A}$$

$$= D \frac{1}{P(x_A) \left(\frac{2\pi}{\beta U''(x_A)} \right)^{\frac{1}{2}}} \frac{e^{\beta U(x_A)} P(x_A)}{e^{\beta U(x^{\ddagger})} \left(\frac{2\pi}{\beta |U''(x^{\ddagger})|} \right)^{\frac{1}{2}}}$$

which simplifies to the celebrated Kramers formula

$$k_A = \frac{\omega_A \omega^{\ddagger}}{2\pi\gamma} e^{-\frac{U^{\ddagger}}{k_B T}} \tag{6.58}$$

where $\omega_A = \sqrt{U''(x_A)}$, $\omega^{\ddagger} = \sqrt{|U''(x^{\ddagger})|}$ are the vibrational frequencies around A and Z, and $U^{\ddagger} = U(x^{\ddagger}) - U(x_A)$ is the barrier height. Consequently, the *Kramers time* or mean first passage time which is the inverse of the transition rate is given by

$$\tau_K = \frac{2\pi\gamma}{\omega_A \omega^{\ddagger}} e^{\beta U^{\ddagger}} \tag{6.59}$$

These equations (6.58) and (6.59) show that the escape rate from A is proportional to the Arrhenius factor $e^{-\frac{U^{\ddagger}}{k_B T}}$, to the vibrational frequencies in the well and at the top of the barrier, and inversely proportional to the viscosity. Similarly, the Kramers time is exponential in the barrier height and in the inverse temperature, proportional to the viscosity of the system and inversely proportional to the vibrational frequencies at the bottom and at the top of the barrier. This formula is verified experimentally in a very wide range of experiments [3].

6.2.2 Mean First Passage Time Derivation

We now show how the Kramers formula can be simply derived using the mean first-passage time equation (6.14). In this case, the domain Ω is the line $[-\infty, x_C]$ and the boundary is $S = \{x_C\}$. We integrate Eq. (6.14) and obtain

$$\frac{\partial \tau_1(x)}{\partial x} = -\frac{1}{D} e^{\beta U(x)} \int_{-\infty}^{x} dz\, e^{-\beta U(z)} \tag{6.60}$$

Integrating this equation, with the boundary condition that $\tau(x_C) = 0$, we have

$$\tau_1(x) = -\frac{1}{D} \int_{x_C}^{x} dy\, e^{\beta U(y)} \int_{-\infty}^{y} dz\, e^{-\beta U(z)} \tag{6.61}$$

and since x is in the domain Ω, $x < x_C$ and we write (6.61) as

$$\tau_1(x) = \frac{1}{D} \int_{x}^{x_C} dy\, e^{\beta U(y)} \int_{-\infty}^{y} dz\, e^{-\beta U(z)} \tag{6.62}$$

We assume that the point x is in the vicinity of A, so that the first passage time τ_K from A to C is given by

$$\tau_K = \frac{1}{D} \int_{x_A}^{x_C} dy\, e^{\beta U(y)} \int_{-\infty}^{y} dz\, e^{-\beta U(z)} \tag{6.63}$$

Assuming again that the barrier is high, we expand both exponentials to second order. In the first integral of $e^{\beta U(y)}$, the exponent is maximal at $x^{\ddagger}$ and we use Eq. (6.56). In the second integral of $e^{-\beta U(z)}$, the exponent is minimal around x_A and we use Eq. (6.54). Since the two exponentials decay rapidly, we extend the bounds of the integrals to $\pm\infty$ and obtain the result of Eq. (6.59)

$$\tau_K = \frac{2\pi\gamma}{\omega_A \omega^{\ddagger}} e^{\beta U^{\ddagger}} \tag{6.64}$$

In Figure 1.2, the Kramers time is the average time the particle spends in one of the wells (say around -1) before jumping to the other well (around $+1$). This time increases exponentially with inverse temperature β and energy barrier $U^{\ddagger}$.

6.3 Transition Path Time: Average and Distribution

We now turn to the specific part of the path where the particle jumps over the barrier from one well to the other [4]. In the same spirit as in the previous section, we model the potential in the vicinity of the barrier Z as a parabola (see Figure 6.2). The barrier potential is thus

$$U(x) = -\frac{K}{2}(x^2 - x_0^2) \tag{6.65}$$

The points D and E, defined on Figure 6.2 denote respectively the origin at $-x_0$ and the extremity at $+x_0$ of the transition path, and Z is the top of the barrier. The specific value of x_0 is somewhat arbitrary. It is chosen so that points D and E delimit boundaries between the potential well around A and the free region on the right of C.

Strictly speaking, the transition path should be defined as the path between points D and E, with absorbing boundary conditions at these points.

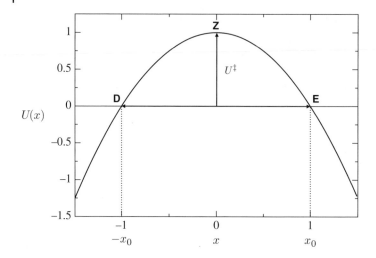

Figure 6.2 Parabolic barrier.

6.3.1 Transition Path Time Distribution

As stated in the previous section, we will show that the transition path time is very short (compared to the Kramers time), and depends weakly on the barrier height (logarithmically). Consider paths going from D to E through the top of the barrier Z. Transition paths are those going from D to E without recrossing D or E. The barrier height is given by

$$U^{\ddagger} = \frac{K}{2}x_0^2 \tag{6.66}$$

If the barrier is high, $U^{\ddagger} \gg k_B T$, once a particle crosses from D to E, the probability of recrossing back to D is very small. In fact, we choose x_0 large enough so that at E, the force at x_0, $-U'(x_0) = Kx_0$ prevents the particle from recrossing back to D. If we assume this, the absorbing boundary condition at E is automatically satisfied, since any particle that crosses E never goes back to D.

Therefore if we neglect recrossing, all paths that start at $-x_0$ at time 0 and end at any point $x > x_0$ at time t, have a transition path time smaller than t. Similarly, paths which start at $-x_0$ at time 0 and have a transition path time shorter than t, necessarily end at some point $x < x_0$ at time t. Therefore we can write

$$\langle \theta(t - t_{TPT}) \rangle = \int_{x_0}^{+\infty} dx \, P(x, t| - x_0, 0) \tag{6.67}$$

and the p.d.f $P_{TPT}(t)$ for the transition path time is given by

$$P_{TPT}(t) = \langle \delta(t - t_{TPT}) \rangle$$

$$= \frac{d}{dt} \int_{x_0}^{+\infty} dx \, P(x, t| - x_0, 0) \tag{6.68}$$

The Langevin Equation for the particle is

$$\frac{dx}{dt} = \omega x + \zeta(t) \tag{6.69}$$

with $\omega = KD\beta$ and $\zeta(t)$ is a white Gaussian noise satisfying the relations

$$\langle \zeta(t) \rangle = 0$$

$$\langle \zeta(t)\zeta(t') \rangle = 2D\delta(t - t') \tag{6.70}$$

The initial condition is $x(0) = -x_0$. Eq. (6.69) can be solved analytically to yield

$$x(t) = -x_0 e^{\omega t} + \int_0^t d\tau e^{\omega(t-\tau)} \zeta(\tau) \tag{6.71}$$

The function $x(t)$ is a linear combination of Gaussian variables $\zeta(\tau)$ and therefore it is Gaussian distributed (see Appendix A). Its distribution is entirely determined by the average and variance of the random variable $x(t)$. We have

$$\langle x(t) \rangle = -x_0 e^{\omega t} \tag{6.72}$$

and the variance is given by

$$\langle (x(t) - \langle x(t) \rangle)^2 \rangle = \int_0^t d\tau \int_0^t d\tau' e^{\omega(2t-\tau-\tau')} \langle \zeta(\tau)\zeta(\tau') \rangle$$

$$= 2D \int_0^t d\tau e^{2\omega(t-\tau)}$$

$$= \frac{1}{K\beta}(e^{2\omega t} - 1)$$

from which we deduce

$$P(x, t| - x_0, 0) = \left(\frac{K\beta}{2\pi(e^{2\omega t} - 1)} \right)^{1/2} \exp\left(-\frac{K\beta(x + x_0 e^{\omega t})^2}{2(e^{2\omega t} - 1)} \right) \tag{6.73}$$

Using Eq. (6.67) we obtain

$$\langle \theta(t - t_{TPT}) \rangle = \int_{x_0}^{+\infty} dx \left(\frac{K\beta}{2\pi(e^{2\omega t} - 1)} \right)^{1/2} \exp\left(-\frac{K\beta(x + x_0 e^{\omega t})^2}{2(e^{2\omega t} - 1)} \right)$$

$$= \frac{1}{2}\left(1 - \text{Erf}\left(x_0 \left(\frac{K\beta}{2} \left(\frac{e^{\omega t} + 1}{e^{\omega t} - 1} \right) \right)^{1/2} \right) \right)$$

$$= \frac{1}{2}\left(1 - \text{Erf}\left(\left(\beta U^{\ddagger} \left(\frac{e^{\omega t} + 1}{e^{\omega t} - 1} \right) \right)^{1/2} \right) \right) \tag{6.74}$$

where we have used the relation (6.66) between x_0 and the energy barrier and the Erf function is defined in Eq. (6.31). In the limit of large time, $t \to \infty$, the above quantity does not go to 1 as one would expect for a distribution function, but rather to $\frac{1}{2}(1 - \text{Erf}\ ((\beta U^{\ddagger})^{1/2}))$. This is due to the fact that some paths originating from D at $-x_0$ go to $-\infty$ and never reach point E. The fraction of paths that go through E at least once in any time is precisely given by this factor $\frac{1}{2}(1 - \text{Erf}\ ((\beta U^{\ddagger})^{1/2}))$. Therefore, in order to normalize the transition path time p.d.f., it is necessary to divide the expression (6.68) by this normalization factor. Taking a derivative of (6.74) with respect to time according to Eq. (6.68), and applying the normalization, we obtain

$$P_{TPT}(t) = \frac{\mathcal{N}}{(1 - e^{-2\omega t})^{1/2}} \frac{e^{-\omega t}}{1 - e^{-\omega t}} e^{-\beta U^{\ddagger}\left(\frac{1+e^{-\omega t}}{1-e^{-\omega t}}\right)} \tag{6.75}$$

where

$$\mathcal{N} = \frac{2\omega}{(1 - \text{Erf}\ ((\beta U^{\ddagger})^{1/2}))} \left(\frac{\beta U^{\ddagger}}{\pi}\right)^{1/2} \tag{6.76}$$

The behaviour of this p.d.f can be easily studied:

1. For a short time $\omega t \ll 1$, we have

$$P_{TPT}(t) \sim \frac{1}{t^{3/2}} e^{-\frac{2\beta U^{\ddagger}}{\omega t}} \tag{6.77}$$

   which has an essential singularity at the origin.
2. For the majority of time $\omega t \gg 1$, it decays exponentially

$$P_{TPT}(t) \sim e^{-\omega t} e^{-\beta U^{\ddagger}} \tag{6.78}$$

In Figure 6.3, we plot the transition path time distribution for various barrier heights. We see the fast decay around 0, and the exponential decay at large times. In recent experiments on protein folding and DNA loop formation, the transition path time distribution has been measured and can be nicely fitted to the form (6.75) [5].

6.3.2 Mean Transition Path Time

As can be seen on Figure 6.3, the transition path time distribution is quite narrow and can be characterized by its mean. By definition, the mean TPT is given by

$$t_{TPT} = \int_0^{+\infty} dt\ t\ P_{TPT}(t) \tag{6.79}$$

In fact, the normalized distribution function $Y(t)$ for the TPT is easily obtained from (6.74) as

$$Y(t) = \frac{1 - \text{Erf}\ (G(t))}{1 - \text{Erf}\ (\sqrt{\beta U^{\ddagger}})} \tag{6.80}$$

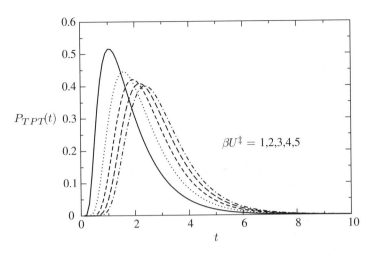

Figure 6.3 Transition Path Time distributions for various barrier heights.

where

$$G(t) = \left(\beta U^{\ddagger} \left(\frac{e^{\omega t} + 1}{e^{\omega t} - 1} \right) \right)^{1/2} \tag{6.81}$$

Taking a derivative of Eq. (6.80), and using the derivative of the Erf function

$$\frac{d}{dx} \text{Erf}(x) = \frac{2}{\sqrt{\pi}} e^{-x^2} \tag{6.82}$$

we have

$$P_{TPT}(t) = -\frac{2}{\sqrt{\pi}} \frac{G'(t) e^{-G^2(t)}}{1 - \text{Erf}(\sqrt{\beta U^{\ddagger}})} \tag{6.83}$$

Using Eq. (6.81), we may express t as a function of G

$$t = -\frac{1}{\omega} \log \left(\frac{G^2 - \beta U^{\ddagger}}{G^2 + \beta U^{\ddagger}} \right) \tag{6.84}$$

Making the change of variable $x = G^2 - \beta U^{\ddagger}$ in (6.79), we have

$$t_{TPT} = -\frac{1}{\omega} \frac{\int_0^{+\infty} \frac{dx}{\sqrt{1 + \frac{x}{\beta U^{\ddagger}}}} \left(\log x - \log \beta U^{\ddagger} - \log 2 - \log \left(1 + \frac{x}{2\beta U^{\ddagger}} \right) \right) e^{-x}}{\int_0^{+\infty} \frac{dx}{\sqrt{1 + \frac{x}{\beta U^{\ddagger}}}} e^{-x}} \tag{6.85}$$

Expanding this expression for large barrier $\beta U^{\ddagger} \gg 1$, we obtain

$$t_{TPT} \simeq \frac{1}{\omega} \log(2 e^C \beta U^{\ddagger}) + O\left(\frac{1}{\beta U^{\ddagger}} \right) \tag{6.86}$$

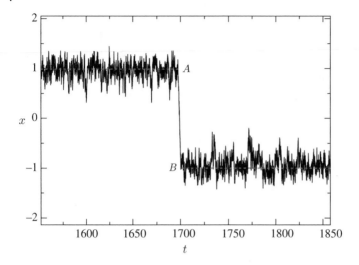

Figure 6.4 Almost ballistic part of the transition path between A and B.

where $C = -\int_0^{+\infty} dx e^{-x} \log x \approx 0.577215$ is the Euler-Mascheroni constant, and the symbol $O\left(\frac{1}{\beta U^{\ddagger}}\right)$ indicates a correction term of order $\frac{1}{\beta U^{\ddagger}}$. This formula was first derived by A.Szabo, unpublished.

As announced in the beginning of this section, the mean TPT is logarithmic in the barrier height, contrary to the Kramers time which is exponential. In many problems of barrier crossing, due to this logarithmic scaling, we have a sharp separation of time scales

$$t_{TPT} \ll t_K \tag{6.87}$$

This relation shows that the transition paths are essentially ballistic, as can be seen on Figure 6.4.

This time scale separation will be used in the next section to devise simple approximation schemes to sample transition paths.

References

1 Redner, S. (2001). *A Guide to First-Passage Processes*. Cambridge University Press.

2 Hänggi, P., Talkner, P., and Borkovec, M. (1990). Reaction-rate theory: fifty years after Kramers, *Rev. Mod. Phys.* 62 (2): 251.

3 (a) Chandler, D. (1978). Statistical mechanics of isomerization dynamics in liquids and the transition state approximation, *J. Chem. Phys.* 68: 2959.

(b) Zhou, H.-X. (2010 May). Rate theories for biologists, *Q. Rev. Biophys.* 43 (2): 219.

4 Zhang, B.W., Jasnow, D., and Zuckerman, D.M. (2007). Transition-event durations in one-dimensional activated processes, *J. Chem. Phys.* 126: 074504. Laleman, M., Carlon, E., and Orland, H. (2017). Transition path time distributions, *J. Chem. Phys.* 147: 214103.

5 Neupane, K., Ritchie, D.B., Yu, H. et al. (2012). Transition path times for nucleic acid folding determined from energy-landscape analysis of single-molecule trajectories, *Phys. Rev. Lett.* 109: 068102.

7

Sampling Transition Paths

In the previous sections, we have very schematically characterized the stochastic trajectories which transport a system from a state A to a state B. Using the terminology of protein folding, if we define A as the unfolded state and B as the folded state, these trajectories are the so-called folding trajectories. The folding time of a protein is directly related to the Kramers time, which is exponential in the free energy barrier height, and ranges from milliseconds to seconds. On the other hand, the timescale which characterizes the barrier crossing from the unfolded state to the folded state is the transition path time. It is logarithmic in the free energy barrier, and is therefore much shorter than the folding time, in the range of microseconds or lower. Looking at barrier crossing trajectories in a simple system (one-dimensional double-well), one sees that during most of the folding time, the system is just sampling the bottom of the well in which it is located. The system is subject to a local harmonic potential and the physics in that region is quite straightforward. The folding effectively takes place during the transition path, over a much shorter time, during which large motion of the system occurs. Given the high complexity of large scale computations and their enormous cost, it is very appealing to try to simulate in an unbiased way, the transition period, so as to be able to understand the mechanisms underlying the folding process. In this section, we will present three methods that can be used to simulate transition paths.

The first idea would be to perform brute force simulations of the system, by discretizing for instance the under- or overdamped Langevin equation, for the protein and for the solvent molecules, until one or several folding events occur. This is a very big computation. Indeed, for a moderately large protein of say 150 aminoacids, representing typically 1500 atoms, one needs also to simulate the water molecules. Given the size of the protein, the simulation must be performed in a box of lateral size of at least 200Å. Such a volume would contain up to 10^5 water molecules. This corresponds to huge simulations, and at present, the best simulations involve volumes containing of the order of 10^4 molecules. It is thus very desirable to be able to perform much shorter simulations, which would typically last the duration of the transition path time. For instance, in the

Molecular Kinetics in Condensed Phases: Theory, Simulation, and Analysis,
First Edition. Ron Elber, Dmitrii E. Makarov and Henri Orland.
© 2020 John Wiley & Sons Ltd. Published 2020 by John Wiley & Sons Ltd.

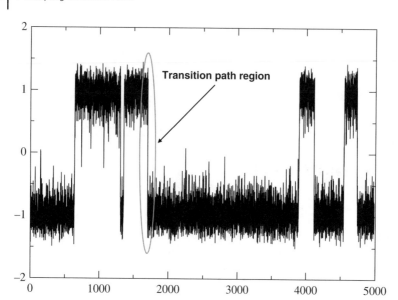

Figure 7.1 In red, the transition path region.

case of the double well potential discussed in the beginning of this chapter, it would be sufficient to study the transition path region in Figure 7.1.

A proper simulation of the transition path should comprise a short sampling period in one of the wells, before the transition, followed by a short sampling period in the second well after the transition. We now proceed to describe three methods to simulate such transition paths.

The idea is to devise an algorithm which generates paths starting in state A at time 0 and ending in state B at time t_f. These path should be unbiased, i.e. they should follow the statistics generated by the Langevin dynamics. Provided t_f is large enough, that is larger than the typical transition path time as defined in the previous section, the transition part of these paths can be identified in the following way. As mentioned above, if t_f is large enough, the path should comprise a first part of quasi-harmonic sampling of the basin of attraction of A, then the transition region, and then again a quasi-harmonic sampling of the basin of attraction of B as in Figure (7.2), which is an enlargement of the transition path part of Figure (7.1). On the other hand, if t_f is smaller than the transition path time, the quasi-harmonic sampling parts are absent, and the path becomes totally determined by the boundary conditions.

In order to study transition paths, we will use the probability $P(x_f, t_f | x_i, 0)$ for the system to start in state A with coordinate x_i at time 0 and to be in state B with coordinate x_f at time t_f. The path integral representation of Eq. (5.6) and

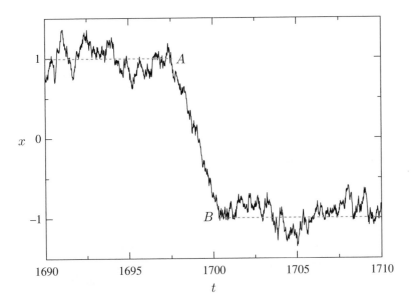

Figure 7.2 Transition path between A and B.

(5.14), can be written in the form

$$P(x_f, t_f | x_i, 0) = \lim_{N \to \infty} \frac{1}{C} \int dx_1 \dots dx_{N-1} e^{-\frac{\gamma dt}{4k_B T} \sum_{n=0}^{N-1} \left(\frac{x_{n+1}-x_n}{dt} + \frac{1}{\gamma} \frac{\partial U}{\partial x}(x_n, \tau_n) \right)^2} \tag{7.1}$$

$$= e^{-\frac{\beta}{2}(U(x_f)-U(x_i))} \int_{x(0)=x_i}^{x(t_f)=x_f} \mathscr{D}x(\tau) \, e^{-\int_0^{t_f} d\tau \left(\frac{1}{4D} \left(\frac{dx}{d\tau} \right)^2 + DV(x(\tau)) \right)} \tag{7.2}$$

$$= e^{-\frac{\beta}{2}(U(x_f)-U(x_i))} \int_{x(0)=x_i}^{x(t_f)=x_f} \mathscr{D}x(\tau) \, e^{-\frac{1}{4k_B T} \int_0^{t_f} d\tau \left(\gamma \left(\frac{dx}{d\tau} \right)^2 + \frac{1}{\gamma} \left(\left(\frac{\partial U(x(\tau))}{\partial x} \right)^2 - 2k_B T \frac{\partial^2 U(x(\tau))}{\partial x^2} \right) \right)} \tag{7.3}$$

where $V(x) = \left(\frac{\beta}{2} \frac{\partial U}{\partial x} \right)^2 - \frac{\beta}{2} \frac{\partial^2 U}{\partial x^2}$ is the potential defined in (3.5) and C is a normalization constant. The boundary conditions are

$$x_0 = x_i$$
$$x_N = x_f \tag{7.4}$$

We now proceed to present some algorithms which allow to sample the most relevant paths contributing to $P(x_f, t_f | x_i, 0)$.

7.1 Dominant Paths and Instantons

Equations (7.1), (7.2) and (7.3) provide a representation of the probability $P(x_f, t_f | x_i, 0)$ as a sum over discrete or continuous paths, of the exponential of a certain function. The first idea that comes to mind is to look for the paths which have the largest contribution to the integrand. The underlying method is the so-called *saddle-point method* (SPM), or *Laplace method*.

7.1.1 Saddle-Point Method

The idea of the method is the following: consider an integral of the form

$$I(l) = \int dx e^{-\frac{1}{l}F(x)} \tag{7.5}$$

and assume we are looking for an asymptotic expansion of $I(l)$ when $l \to 0$. The factor $1/l$ in front of $F(x)$ is very large and thus selects the vicinity of the minimum of $F(x)$. We are thus led to expand $F(x)$ around its minimum x_0. The minimum is defined by

$$F'(x_0) = 0$$
$$F''(x_0) > 0 \tag{7.6}$$

and we expand F to second order as

$$F(x) = F(x_0) + \frac{1}{2}(x - x_0)^2 F''(x_0) + \dots \tag{7.7}$$

Replacing Eq. (7.7) in (7.5), we have

$$I(l) = \int dx e^{-\frac{1}{l}\left(F(x_0) + \frac{1}{2}(x-x_0)^2 F''(x_0) + \dots\right)}$$
$$\approx_{l \to 0} \frac{e^{-F(x_0)/l}}{\sqrt{2\pi l F''(x_0)}} \tag{7.8}$$

The dominant term in Eq. (7.8) is the exponential term, and the first correction, due to the quadratic fluctuations of $F(x_0)$ in Eq. (7.7), is the square root denominator. Expanding $F(x)$ in Eq. (7.7) to higher order around x_0 and performing the integrals provide an asymptotic expansion of the logarithm of $I(l)$ in powers of $1/l$.

7.1.2 The Euler-Lagrange Equation: Dominant Paths

This approach is similar in essence to the *semi-classical* or *WKB* approximation of quantum mechanics ([1]). In the following, we will restrict ourselves to the search of the minimal transition path which connects the two extremities. Defining the *Lagrangian* function

$$L(x(t)) = \frac{\gamma}{4}\dot{x}^2 + k_B T D V(x)$$

$$= \frac{\gamma}{4}\dot{x}^2 + \frac{1}{4\gamma}\left(\left(\frac{\partial U(x(\tau))}{\partial x}\right)^2 - 2k_B T \frac{\partial^2 U(x(\tau))}{\partial x^2}\right) \quad (7.9)$$

the probability takes the form

$$P(x_f, t_f | x_i, 0) = e^{-\frac{\beta}{2}(U(x_f) - U(x_i))} \int_{x(0)=x_i}^{x(t_f)=x_f} \mathcal{D}x(\tau) \, e^{-\frac{1}{k_B T}\int_0^{t_f} d\tau L(x(\tau))} \quad (7.10)$$

By definition, the *action S* is the integral of the Lagrangian

$$S = \int_0^{t_f} d\tau L(x(\tau)) \quad (7.11)$$

In the following, we will assume that the friction coefficient γ does not depend on temperature. In Eq. (7.10), the exponent is proportional to $1/T$ and thus at low temperature, $T \to 0$, we may want to use the SPM described above. Note however that the Lagrangian (7.9) depends on T through the second derivative term, and thus the SPM will not provide an exact expansion in powers of $1/T$. However, it can be shown that the dominant order is correct. The paths which minimize $\int_0^{t_f} d\tau L(x(\tau))$ and satisfy the boundary conditions $x(0) = x_i$ and $x(t_f) = x_f$ are those which provide the dominant contribution to $P(x_f, t_f | x_i, 0)$ at low temperature. As is well known from classical Lagrangian mechanics, these paths are given by the *Euler-Lagrange equations* (EL). The general form of these equations is

$$\frac{d}{dt}\left(\frac{\partial L}{\partial \dot{x}}\right) = \frac{\partial L}{\partial x} \quad (7.12)$$

In the present case, the EL equations become

$$\frac{d^2x}{dt^2} = 2D^2 \frac{\partial}{\partial x} V(x)$$

$$= \frac{1}{\gamma^2}\frac{\partial}{\partial x}\left(\frac{1}{2}\left(\frac{\partial U(x(t))}{\partial x}\right)^2 - k_B T \frac{\partial^2 U(x(t))}{\partial x^2}\right) \quad (7.13)$$

to be solved with the boundary conditions

$$x(0) = x_i$$

$$x(t_f) = x_f \quad (7.14)$$

These paths correspond to the so-called *classical paths* in quantum mechanics, and are called *instantons* when they connect local minima of the energy

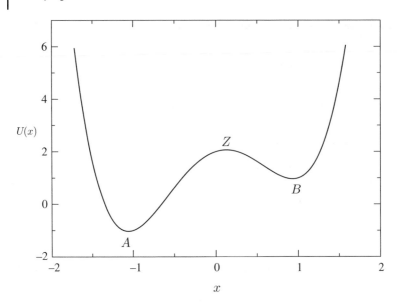

Figure 7.3 Asymmetric quartic potential. The points A and B are minima, and Z is a maximum.

$U(x)$ [2][1]. In the framework of Langevin dynamics, they are called *dominant paths*, since they have maximum contribution to P. Had we started from the discrete form of the path integral (7.1), the dominant paths would have been those minimizing the discrete weight $\sum_{n=0}^{N-1} \left(\frac{x_{n+1} - x_n}{dt} + \frac{1}{\gamma} \frac{\partial U}{\partial x}(x_n, \tau_n) \right)^2$. The minimization leads to an equation which in the continuous limit is identical to (7.13) but without the second derivative term of the potential V.

One can identify Eq. (7.13) as the Newton equation for a classical particle in the effective potential $-2D^2 V(x) = \frac{1}{\gamma^2} \left(k_B T \frac{\partial^2 U(x(t))}{\partial x^2} - \frac{1}{2} \left(\frac{\partial U(x(\tau))}{\partial x} \right)^2 \right)$. At low temperature, the square term is dominant and the effective potential is negative. It is maximum close to the minima and maxima of $U(x)$ (see Figures (7.3, 7.4)). Multiplying Eq. (7.13) by $\dot{x}$ and integrating the equation, we see that there is a quantity conserved by this equation, namely the pseudo-energy $\mathscr{E}$ given by

$$\mathscr{E} = \frac{1}{4D}\dot{x}^2 - DV(x) \tag{7.15}$$

Given the initial and final points x_i and x_f, the above equation implies the following relation between the initial and final velocities

$$\frac{1}{4D}(\dot{x}_f^2 - \dot{x}_i^2) = D(V(x_f) - V(x_i)) \tag{7.16}$$

---

1 Note that if we wanted an exact low temperature expansion, one should discard the second derivative term in Eq. (7.13), since it is of order T.

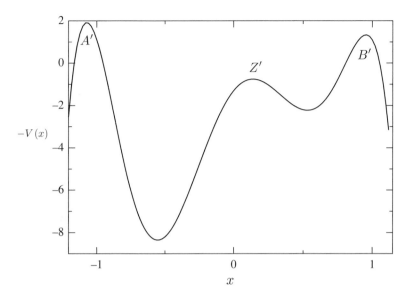

Figure 7.4 Effective potential $-V(x)$ as a function of x. The three points A', Z' and B' are maxima.

If for example $V(x_f) = V(x_i)$, then the initial and final velocities should be identical in absolute value. The conservation of the pseudoenergy allows for the solution of the equation of motion in one dimension. Indeed, using Eq. (7.15), the movement of the particle in the A' to B' region in the forward direction, is given by

$$\dot{x} = \sqrt{4D(\mathscr{E} + DV(x))} \tag{7.17}$$

or

$$dt = \frac{dx}{\sqrt{4D(\mathscr{E} + DV(x))}} \tag{7.18}$$

which can be integrated as

$$t = \int_{x_i}^{x(t)} \frac{dx'}{\sqrt{4D(\mathscr{E} + DV(x'))}} \tag{7.19}$$

This equation reduces the motion of a particle in 1d to a simple integration.

In the multidimensional case $U(x_1, \cdots, x_N)$, the EL equations read

$$\frac{d^2 x_k}{dt^2} = \frac{1}{\gamma^2} \frac{\partial}{\partial x_k} \sum_{l=1}^{N} \left(\frac{1}{2} \left(\frac{\partial U(\{x_m(t)\})}{\partial x_l} \right)^2 - k_B T \frac{\partial^2 U(\{x_m(t)\}}{\partial x_l^2} \right) \tag{7.20}$$

for each degree of freedom $x_k(t)$, with the boundary onditions $x_k(0) = x_k^{(i)}$ and $x_k(t_f) = x_k^{(f)}$ for all k.

These second order differential equations, together with the boundary conditions, determine the paths which contribute maximally to the transition probability. These equations (7.13) or (7.20) are solved by discretization. The difficulty lies in enforcing the boundary conditions. In addition, there may be several dominant paths $x_S(t)$ satisfying the boundary conditions. The conserved pseudoenergy is

$$\mathcal{E} = \sum_{l=1}^{N} \left(\frac{1}{4D}\dot{x}_l^2 - \frac{D\beta^2}{4} \left(\frac{\partial U(\{x_m(t)\})}{\partial x_l} \right)^2 + \frac{D\beta}{2} \frac{\partial^2 U(\{x_m(t)\})}{\partial x_l^2} \right) \quad (7.21)$$

but its conservation is not sufficient to solve for the trajectory in dimensions larger than 1.

Several techniques can be used to solve these equations, such as the shooting method, relaxations methods, etc. In general, it is a difficult problem in high dimensions, and we refer the reader to the mathematical literature for details on the technical implementation.

7.1.3 Steepest Descent Method

We briefly summarize here a simple *steepest descent method* to determine dominant paths. Consider the discretized form of the integral of the Lagrangian in Eq. (7.10), where the timestep dt must be chosen small enough to obtain meaningful results. We look for minima of the exponent

$$S = \sum_{n=0}^{N-1} \left(\frac{1}{4D} \left(\frac{x_{n+1} - x_n}{dt} \right)^2 + DV(x_n) \right) dt \quad (7.22)$$

We start from an initial path $\{x_n^{(0)}\}$ satisfying the boundary conditions (7.4). The corresponding exponent (7.22) is denoted $S^{(0)}$. We now proceed to deform this path by displacing each point $x_n^{(0)}$ without modifying its extremities, and accepting the displacement only if it decreases the action $S^{(0)}$. Practically, we can scan each point $x_n^{(0)}$ of the trajectory sequentially or in random order. We try to replace $x_n^{(0)}$ by $x_n^{(0)} + \delta x_n$ where δx_n is some small enough increment (see Figure (7.5)), which will be adjusted in the process. There are only two terms in S which contain $x_n^{(0)}$ and the variation δS_n of the exponent is thus

$$\delta S_n = \frac{1}{4D} \left(\frac{x_{n+1}^{(0)} - x_n^{(0)} - \delta x_n}{dt} \right)^2 + \frac{1}{4D} \left(\frac{x_n^{(0)} + \delta x_n - x_{n-1}^{(0)}}{dt} \right)^2 + DV(x_n^{(0)} + \delta x_n)$$

$$- \frac{1}{4D} \left(\frac{x_{n+1}^{(0)} - x_n^{(0)}}{dt} \right)^2 - \frac{1}{4D} \left(\frac{x_n^{(0)} - x_{n-1}^{(0)}}{dt} \right)^2 - DV(x_n^{(0)}) \quad (7.23)$$

If $\delta S_n \leq 0$, the displacement of $x_n^{(0)}$ is accepted and the new variable is $x_n^{(1)} = x_n^{(0)} + \delta x_n$, otherwise if $\delta S_n > 0$, the variation is rejected and $x_n^{(1)} = x_n^{(0)}$.

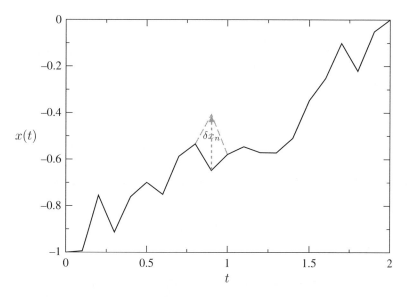

Figure 7.5 Illustration of the update of a trajectory at a given time.

We thus have a new trajectory and the procedure is iterated until all variations are rejected. At that stage, the increments δx_n must be decreased by some factor, and one should restart the whole procedure, until the increments δx_n become smaller than a predefined threshold which defines convergence. At the end of this procedure, the trajectory has been deformed into the closest local minimum of the exponent (7.22). As mentioned above, there may be several dominant trajectories, and they may be explored by renewing the above process with a different initial trajectory.

7.1.4 Gradient Descent Method

A variant of this method is the *gradient descent* method. The idea is again to start from an initial trajectory $\{x_n^{(0)}\}$ which satisfies the boundary conditions (7.4), and to solve the recursion relation

$$x_n^{(k+1)} = x_n^{(k)} - \lambda \frac{\partial S}{\partial x_n^{(k)}} \tag{7.24}$$

where λ is a positive parameter to be determined and k is an iteration index. One can show that for small enough λ, the exponent S necessarily decreases at each iteration, until convergence. The criterion for determining λ is thus to choose the largest value which still allows for a decrease of S. Both methods described above can be easily generalized to the case of many degrees of freedom. However there is no guarantee that they will provide all the existing dominant paths, let alone the most important ones.

As mentioned above, there may be several dominant paths $x_S(t)$ joining the extremities. Each path $x_S(t)$ has a weight given by Eq. (7.3). and to lowest approximation, we have

$$P(x_f, t_f | x_i, 0) = e^{-\frac{\beta}{2}(U(x_f) - U(x_i))} \sum_{\text{all dominant paths}} e^{-\frac{1}{k_B T} \int_0^{t_f} d\tau L(x_S(\tau))} \tag{7.25}$$

Since dominant paths are weighted by the factor $1/k_B T$, the ones with the lowest action are exponentially dominant compared to the others. However, as noted above, there may be many dominant paths and it is not clear how to find those with least action.

7.2 Path Sampling

Instead of looking for the dominant paths, it is appealing to sample directly the paths. For that matter, we can easily transform the steepest descent method into a Metropolis scheme, and the gradient descent method into a Langevin scheme.

7.2.1 Metropolis Scheme

The strategy is very similar to that of the steepest descent method:

1. Start with an initial trajectory $\{x_n^{(0)}\}$ constructed by some method with assigned endpoints (7.4).
2. Pick randomly an index $n \in \{1, \dots N\}$. Try to replace $x_n^{(0)}$ by $x_n^{(0)} + \delta x_n$ where δx_n is a small enough increment.
3. Compute the variation of exponent δS_n given by Eq. (7.23)
4. Compute $p = e^{-\delta S_n}$ as prescribed by the path integral (7.1). Accept the update of $x_n^{(0)}$ to $x_n^{(0)} + \delta x_n$ with probability p, i.e.
   a) if $p \geq 1$, accept the move: $x_n = x_n^{(0)} + \delta x_n$
   b) if $p \leq 1$, draw uniformly a random number r in the interval $[0, 1]$. If $r \leq p$, accept the move, if $r > p$, reject the move
5. Back to 2.

The magnitude of the increments δx_n is chosen so that the acceptance rate for the changes of x_n is close to $1/2$. Indeed, if the acceptance rate is close to 1, it indicates that δx_n is so small that almost all moves are accepted, whereas a small acceptance rate is a sign that δx_n is too large and consequently, most moves are rejected. The δx_n are determined by performing some preliminary runs of adjustment, and monitoring the acceptance rate. Once the increments are properly adjusted, it can be shown that iterating this algorithm over long enough time, provides a statistical sample of the space of trajectories joining x_i to x_f during time t_f, properly weighted by the Onsager-Machlup action.

7.2.2 Langevin Scheme

Here, the strategy is similar to that of the gradient descent method. The idea is to introduce a fictitious time τ and evolve the whole trajectory in time with a Langevin dynamics, so that at stationarity, a statistical sample is constructed. The time τ is discretized, with running index k and timestep δt. Starting from an initial trajectory $\{x_n^{(0)}\}$ satisfying the boundary conditions (7.4), we iterate the analogous to Eq. (7.24), adding noise to it

$$x_n^{(k+1)} = x_n^{(k)} - \delta t \frac{\partial S}{\partial x_n^{(k)}} + \sqrt{2\delta t} \xi_n^{(k)} \tag{7.26}$$

where the $\xi_n^{(k)}$ are identically distributed Gaussian variables with zero mean and variance 1, and δt is a small parameter, adjusted so that the algorithm does not become unstable. Again, in the long time limit, the trajectories generated by this method sample a statistical ensemble where the weight is given by (7.22).

Both methods can easily be generalized to the case of many degrees of freedom. It carries some of the drawbacks of stochastic sampling methods, namely the difficulty to overcome high barriers. Indeed, there may be many valleys of paths, separated by high barriers. Direct sampling as described above may restrict the search to the valleys where the initial path was picked from, or to their vicinity, and produce samples of paths strongly correlated with the initial choice. To overcome this difficulty, one may use *simulated annealing* [3], i.e. perform the sampling at a high temperature, and slowly lower it until it reaches the actual temperature. But this method is computationally very heavy and intensive with no guarantee of eventually sampling the most relevant trajectories.

7.3 Bridge and Conditioning

As discussed in the preceding sections, the main drawback of the methods proposed above is that they provide families of strongly correlated trajectories due to the high barriers separating the different valleys. A way around this is to use so-called *bridge equations*. The bridge equations which will discussed here were first introduced by Doob [4]. The idea is to write a modified Langevin Equation, with an additional term which conditions the particle to end at x_f at time t_f, without introducing any bias into the statistics. Indeed, among the infinite set of Langevin trajectories starting at x_i at time 0, for any given time t_f, there is a subset of trajectories which ends at x_f at t_f. The idea of the bridge equations is to generate only these trajectories, with the right statistics.

Consider the set of all trajectories starting at x_i at $t_i = 0$ and ending at x_f at time t_f. What is the probability $\mathscr{P}(x, t)$ to find a particle at point x at intermediate time t? It is easy to see that this conditional probability is given by

$$\mathscr{P}(x, t) = \frac{P(x_f, t_f | x, t) P(x, t | x_i, 0)}{P(x_f, t_f | x_i, 0)} \tag{7.27}$$

The probability for a particle going from $(x_i, 0)$ to (x_f, t_f) to be at (x, t), is equal to the probability that the particle goes first from $(x_i, 0)$ to (x, t), and then from (x, t) to (x_f, t_f). This probability should be properly normalized. In fact, due to the Chapman–Kolmogorov equation

$$P(x_f, t_f | x_i, 0) = \int dx\, P(x_f, t_f | x, t) P(x, t | x_i, 0) \tag{7.28}$$

we have

$$\int dx \mathscr{P}(x, t) = 1 \tag{7.29}$$

From section 2, we know that $P(x, t | x_i, 0)$ satisfies the Fokker–Planck equation (2.17)

$$\frac{dP(x, t | x_i, 0)}{dt} = D \frac{\partial}{\partial x} \left(\frac{\partial P(x, t | x_i, 0)}{\partial x} - \beta F(x) P(x, t | x_i, 0) \right) \tag{7.30}$$

and $P(x_f, t_f | x, t)$ satisfies the adjoint Fokker–Planck equation (2.32)

$$\frac{dP(x_f, t_f | x, t)}{d\tau} = -D \left(\frac{\partial^2 P(x_f, t_f | x, t)}{\partial x^2} + \beta F(x) \frac{\partial P(x_f, t_f | x, t)}{\partial x} \right) \tag{7.31}$$

Taking a derivative with respect to time of equation (7.27) and using the above two equations, we have

$$\frac{d\mathscr{P}(x, t)}{dt} = \frac{1}{P(x_f, t_f | x_i, 0)} \left(\frac{dP(x_f, t_f | x, t)}{dt} P(x, t | x_i, 0) + P(x_f, t_f | x, t) \frac{dP(x, t | x_i, 0)}{dt} \right)$$

$$= \frac{D}{P(x_f, t_f | x_i, 0)} \left\{ -\left(\frac{\partial^2 P(x_f, t_f | x, t)}{\partial x^2} + \beta F(x) \frac{\partial P(x_f, t_f | x, t)}{\partial x} \right) P(x, t | x_i, 0) \right.$$

$$\left. + P(x_f, t_f | x, t) \frac{\partial}{\partial x} \left(\frac{\partial P(x, t | x_i, 0)}{\partial x} - \beta F(x) P(x, t | x_i, 0) \right) \right\}$$

After some simple rearrangement of the above, we obtain

$$\frac{d\mathscr{P}(x, t)}{dt} = D \frac{\partial}{\partial x} \left(\frac{\partial \mathscr{P}(x, t)}{\partial x} - \beta \left(F(x) + 2 k_B T \frac{\partial}{\partial x} \ln Q(x, t) \right) \mathscr{P}(x, t) \right) \tag{7.32}$$

where we have introduced the notation

$$Q(x, t) = P(x_f, t_f | x, t) \tag{7.33}$$

Equation (7.32) is remarkable, as it shows that the conditioned probability distribution to find the particle at (x, t) satisfies a modified Fokker–Planck equation. Indeed, Eq. (7.32) is the Fokker–Planck equation for a particle subject to a force $F(x) + 2k_B T \frac{\partial}{\partial x} \ln Q(x, t)$. This equation can thus be viewed as the Fokker–Planck equation derived from the Langevin equation for a particle subject to the same force. We deduce that the motion of a Langevin particle subject to a force $F(x)$ and conditioned to end at point x_f at time t_f is given by

$$\frac{dx}{dt} = D\beta F(x(t)) + 2D \frac{\partial}{\partial x} \ln Q(x, t) + \eta(t) \tag{7.34}$$

where $\eta(t)$ is a Gaussian white noise with moments given by (1.43)

$$\langle \eta(t) \rangle = 0$$
$$\langle \eta(t)\eta(t') \rangle = 2D\delta(t - t') \tag{7.35}$$

This equation (7.34) is called the *bridge equation*. It is exact and generates all trajectories which start at x_i at 0 and end at x_f at time t_f with the correct statistics. There is a guiding force $2k_B T \frac{\partial}{\partial x} \ln Q(x, t)$ which drives the particle to the correct end point x_f at the expected time t_f, without biasing the statistics. It can be shown that this is the only force which has this property. This force is singular at t_f. Indeed, we have

$$Q(x, t_f) = P(x_f, t_f | x, t_f) = \delta(x_f - x) \tag{7.36}$$

and thus Q and all its derivative become singular as $t \to t_f$. It is this strong singularity which drives the particle to its predefined target. Because this force is local in space and time, the bridge equation is Markovian, and all theorems pertaining to Markov processes apply. As usual, this bridge equation can be easily generalized to the case of many degrees of freedom.

In the case when the force is conservative, we can use the identity (3.13) for $Q(x, t)$

$$Q(x, t) = e^{-\frac{\beta}{2}(U(x_f) - U(x))} I(x, t) \tag{7.37}$$

where

$$I(x, t) = \langle x_f | e^{-H(t_f - t)} | x \rangle \tag{7.38}$$

Taking the logarithmic derivative of these equations and replacing in Eq. (7.34), the bridge equation takes the form

$$\frac{dx}{dt} = 2D \frac{\partial}{\partial x} \ln I(x, t) + \eta(t) \tag{7.39}$$

and I satisfies the reverse-time Schrödinger equation

$$\frac{\partial I}{\partial t} = D\left(-\frac{\partial^2}{\partial x^2} + V(x)\right) I(x, t) \tag{7.40}$$

The main difficulty of these equations (7.34) or (7.39), lies in the calculation of the guiding function $Q(x, t)$. In the following, we will show some simple exactly solvable examples, followed by some approximations which are valid when there is clear time scale separation.

7.3.1 Free Particle

We consider a free Brownian particle, and show how to generate sample trajectories starting at x_i at time 0 and ending at x_f at time t_f. The potential U is equal to 0, and thus the function Q can be easily evaluated. Indeed, using the results of section (2.4.1), Eq. (2.43), we have

$$Q(x, t) = P(x_f, t_f | x, t)$$

$$= \frac{1}{\sqrt{4\pi D(t_f - t)}} e^{-\frac{(x_f - x)^2}{4D(t_f - t)}} \tag{7.41}$$

Taking the logarithmic derivative of this expression and inserting in (7.34), we obtain the Brownian bridge equation

$$\frac{dx}{dt} = \frac{x_f - x}{t_f - t} + \eta(t) \tag{7.42}$$

We see that the guiding term diverges as $t \to t_f$, and forces the particle at x to go to x_f. The driving term does not depend on the diffusion constant and is totally universal. This equation can be trivially solved by discretization. In Figure (7.6), we show a sample of such bridge trajectories, starting at -1 at time 0 and ending at $+1$ at time 1.

Other types of exactly solvable bridges, with different boundary conditions, such as half-planes, etc., can be found in [5].

7.3.2 The Ornstein-Uhlenbeck Bridge

Another useful case where the bridge equations can be obtained exactly is that of the Ornstein–Uhlenbeck process, i.e. an overdamped particle in a harmonic potential, see sec. 1.4. The Langevin Equation takes the form of Eq. (1.55)

$$\frac{dx}{dt} = -D\beta Kx + \eta(t)$$

where K is the rigidity constant of the harmonic oscillator, Eq. (1.54). As in the free case, the Green's function $Q(x, t)$ can be computed exactly. In fact, it can be

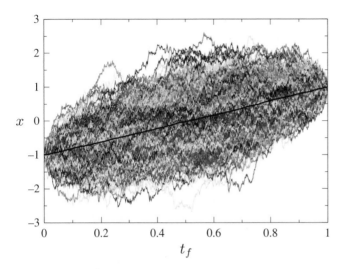

Figure 7.6 Brownian bridges. (*See color plate section for color representation of this figure*).

easily obtained from Eq. (1.59). Indeed, this equation actually gives the expression for $P(x, t|x_0, 0)$. We can obtain $P(x_f, t_f|x, t)$ by making the replacement

$$x \to x_f$$
$$t \to t_f$$
$$x_0 \to x$$
$$t \to t_f - t$$

The result is

$$Q(x, t) = \frac{1}{\mathcal{N}} \exp\left(-\frac{\beta K(x_f - xe^{-\omega(t_f-t)})^2}{2(1 - e^{-2\omega(t_f-t)})}\right) \qquad (7.43)$$

where $\mathcal{N}$ is a normalization constant which depends only on time, and $\omega = D\beta K$. Taking the logarithmic derivative of the above, we obtain the bridge equation for the Ornstein–Uhlenbeck particle

$$\frac{dx}{dt} = \omega\frac{x_f - x\cosh\omega(t_f - t)}{\sinh\omega(t_f - t)} + \eta(t) \qquad (7.44)$$

As in the free case, this equation can be solved by discretization. It is remarkable to note that the conditioned paths do not depend on the sign of the constant K or ω. For instance, on Figure (7.7), trajectories from A to Z or from Z to A are generated by the same equation (7.43).

This means that the trajectories going up a barrier or going down a well are statistically identical. This could already be seen on Figure (7.7) and Figure (1.2) (trajectories in a double well), where the part of the trajectories between A and

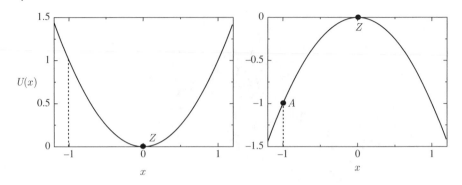

Figure 7.7 Conditioned trajectories from A to Z in a well (left) or over a barrier (right) are statistically identical.

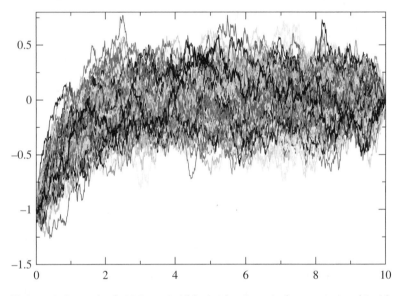

Figure 7.8 A sample of 100 Ornstein-Uhlenbeck trajectories between -1 and 0 with $t_f = 10$. (*See color plate section for color representation of this figure*).

Z, corresponding to a particle falling in the well (from -1 to 0, left figure), is statistically identical to that of the trajectory going up a barrier (from -1 to 0, right figure). In Figure (7.8), we plot a sample of 100 bridge trajectories in a parabolic potential, from -1 to 0, as in the right part of Figure (7.7).

7.3.3 Exact Diagonalization

In general, it is not possible to compute analytically the guiding function $Q(x, t)$. In some cases however, it can be computed numerically exactly. In the following, we will use the notations and definitions of section 3. We recall that from Eqs. (7.37), (7.38) and (3.15) we have Figure 7.8

$$Q(x,t) = e^{-\frac{\beta}{2}(U(x_f)-U(x))}\langle x_f|e^{-H(t_f-t)}|x\rangle \tag{7.45}$$

$$= e^{-\frac{\beta}{2}(U(x_f)-U(x))} \sum_{n=0}^{\infty} e^{-E_n(t_f-t)}\langle x_f|\Psi_n\rangle\langle\Psi_n|x\rangle \tag{7.46}$$

$$= e^{-\frac{\beta}{2}(U(x_f)-U(x))} \sum_{n=0}^{\infty} e^{-E_n(t_f-t)}\Psi_n(x_f)\Psi_n(x) \tag{7.47}$$

where the $|\Psi_n\rangle$ and E_n are the eigenstates and eigenvalues of the Hamiltonian $H = D\left(-\frac{\partial^2}{\partial x^2} + V(x)\right)$ with $V(x) = \left(\frac{\beta}{2}\frac{\partial U}{\partial x}\right)^2 - \frac{\beta}{2}\frac{\partial^2 U}{\partial x^2}$. Taking the logarithmic derivative of the above expressions, we have

$$\frac{\partial}{\partial x}\ln Q(x,t) = \frac{\beta}{2}\frac{\partial U(x)}{\partial x} + \frac{\sum_{n=0}^{\infty} e^{-E_n(t_f-t)}\Psi_n(x_f)\frac{\partial}{\partial x}\Psi_n(x)}{\sum_{n=0}^{\infty} e^{-E_n(t_f-t)}\Psi_n(x_f)\Psi_n(x)} \tag{7.48}$$

so that the bridge equation (7.34) becomes

$$\frac{dx}{dt} = 2D\frac{\sum_{n=0}^{\infty} e^{-E_n(t_f-t)}\Psi_n(x_f)\frac{\partial}{\partial x}\Psi_n(x)}{\sum_{n=0}^{\infty} e^{-E_n(t_f-t)}\Psi_n(x_f)\Psi_n(x)} + \eta(t) \tag{7.49}$$

This exact equation can be generalized to any number of degrees of freedom, and can be used to generate transition paths, provided one can diagonalize the Hamiltonian H. With a small enough number of degrees of freedom (typically up to 5), one can diagonalize numerically the Hamiltonian, and solve Eq. (7.49). For example, for the case of the quartic double well $U(x) = \frac{1}{4}(x^2 - 1)^2$ of Figure (1.3), we can diagonalize numerically the Hamiltonian H. In Figure (7.9), we show a set of bridge trajectories going from one well at $x_i = -1$ to the other well at $x_f = +1$ with $t_f = 100$.

7.3.4 Cumulant Expansion

In general, it is not possible to find an analytic expression for $Q(x,t)$ or even to compute it exactly numerically by diagonalization of the Hamiltonian. Indeed, when studying a typical system, like thermal folding–unfolding of a protein, the system comprises a few thousand degrees of freedom, which makes the

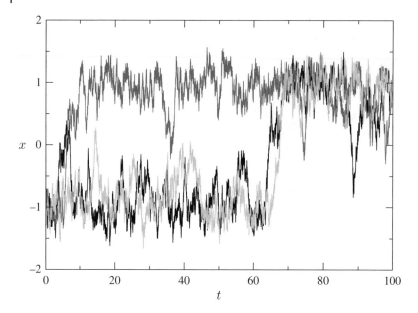

Figure 7.9 Three bridge trajectories in a double well, joining the two local minima. (*See color plate section for color representation of this figure*).

numerical diagonalization totally impossible. One must thus resort to some approximation [6, 7]. In the following, we will consider the problem of barrier crossing, for a multi-particle system. We will be interested in simulating only the transition path portion of the trajectory.

We start from the path-integral representation Eqs. (7.37) and (7.38)

$$Q(x, t) = e^{-\frac{\beta}{2}(U(x_f)-U(x))} I(x, t) \tag{7.50}$$

where

$$I(x, t) = \langle x_f | e^{-H(t_f - t)} | x \rangle = \int_{x(t)=x}^{x(t_f)=x_f} \mathscr{D}x(\tau) \, e^{-\int_t^{t_f} d\tau \left(\frac{1}{4D} \left(\frac{dx}{d\tau} \right)^2 + DV(x(\tau)) \right)} \tag{7.51}$$

We define the quantity $I_0(x, t)$, analogous to $I(x, t)$ but with no potential. Using Eq. (7.41), we have

$$I_0(x, t) = \int_{x(t)=x}^{x(t_f)=x_f} \mathscr{D}x(\tau) \, e^{-\frac{1}{4D}\int_t^{t_f} d\tau \left(\frac{dx}{d\tau} \right)^2} = \frac{1}{\sqrt{4\pi D(t_f - t)}} e^{-\frac{(x_f - x)^2}{4D(t_f - t)}} \tag{7.52}$$

We may thus write

$$
\frac{I(x,t)}{I_0(x,t)} = \frac{\int_{x(t)=x}^{x(t_f)=x_f} \mathcal{D}x(\tau)\, e^{-\frac{1}{4D}\int_t^{t_f} d\tau \left(\frac{dx}{d\tau}\right)^2} e^{-D\int_t^{t_f} d\tau V(x(\tau))}}{\int_{x(t)=x}^{x(t_f)=x_f} \mathcal{D}x(\tau)\, e^{-\frac{1}{4D}\int_t^{t_f} d\tau \left(\frac{dx}{d\tau}\right)^2}}
\tag{7.53}
$$

$$
= \left\langle e^{-D\int_t^{t_f} d\tau V(x(\tau))} \right\rangle_0
\tag{7.54}
$$

where $\langle \dots \rangle_0$ denotes the expectation value with respect to the free particle

$$
\langle A \rangle_0 = \frac{\int_{x(t)=x}^{x(t_f)=x_f} \mathcal{D}x(\tau)\, A(x(\tau))\; e^{-\frac{1}{4D}\int_t^{t_f} d\tau \left(\frac{dx}{d\tau}\right)^2}}{\int_{x(t)=x}^{x(t_f)=x_f} \mathcal{D}x(\tau)\, e^{-\frac{1}{4D}\int_t^{t_f} d\tau \left(\frac{dx}{d\tau}\right)^2}}
\tag{7.55}
$$

We now use the *cumulant expansion*

$$
\langle e^{-A} \rangle_0 \approx 1 - \langle A \rangle_0 + \frac{1}{2}\langle A^2 \rangle_0 + \dots
$$
$$
\approx e^{-\langle A \rangle_0 + \frac{1}{2}(\Delta A)^2 + \dots}
\tag{7.56}
$$

where the variance is given by

$$
(\Delta A)^2 = \langle A^2 \rangle_0 - \langle A \rangle_0^2
\tag{7.57}
$$

If we assume the variance to be small $(\Delta A)^2 \ll \langle A \rangle_0$, we can neglect the fluctuations, and use the approximation

$$
\langle e^{-A} \rangle_0 \approx e^{-\langle A \rangle_0}
\tag{7.58}
$$

In the case of barrier crossing, as we inferred in section (6.3.2), the duration τ_{TPT} of the transition path is very short compared to the escape time τ_K from the well, $\tau_{TPT} \ll \tau_K$, and thus the transition paths are almost ballistic at low temperature, with very little diffusion effects. As a result, the fluctuations in the transition paths are small and we may neglect the variance and use Eq. (7.58). Using Eqs. (7.51) and (7.52), we obtain

$$
I(x,t) = I_0(x,t) \exp\left(-D \int_t^{t_f} d\tau \langle V(x(\tau)) \rangle_0 \right)
\tag{7.59}
$$

where the average is given by

$$
\langle V(x(\tau)) \rangle_0 = \frac{\int_{x(t)=x}^{x(t_f)=x_f} \mathcal{D}x(\tau)\, e^{-\frac{1}{4D}\int_t^{t_f} d\tau \left(\frac{dx}{d\tau}\right)^2} V(x(\tau))}{\int_{x(t)=x}^{x(t_f)=x_f} \mathcal{D}x(\tau)\, e^{-\frac{1}{4D}\int_t^{t_f} d\tau \left(\frac{dx}{d\tau}\right)^2}}
\tag{7.60}
$$

It is convenient to rewrite

$$\int_{x(t)=x}^{x(t_f)=x_f} \mathscr{D}x(\tau)\, e^{-\frac{1}{4D}\int_t^{t_f} d\tau \left(\frac{dx}{d\tau}\right)^2} V(x(\tau))$$

$$= \int_{x(t)=x}^{x(t_f)=x_f} \mathscr{D}x(\tau)\, e^{-\frac{1}{4D}\int_\tau^{t_f} d\tau'\left(\frac{dx}{d\tau'}\right)^2} V(x(\tau)) e^{-\frac{1}{4D}\int_t^\tau d\tau'\left(\frac{dx}{d\tau'}\right)^2}$$

$$= \int dz\, P_0(x_f, t_f | z, \tau) V(z) P_0(z, \tau | x, t)$$

$$= \int dz \frac{1}{\sqrt{4\pi D(t_f - \tau)}} \frac{1}{\sqrt{4\pi D(\tau - t)}} e^{-\frac{(x_f - z)^2}{4D(t_f - \tau)}} e^{-\frac{(z-x)^2}{4D(\tau - t)}} V(z)$$

We have

$$\langle V(x(\tau))\rangle_0 = \frac{\int dz \dfrac{1}{\sqrt{4\pi D(t_f - \tau)}} \dfrac{1}{\sqrt{4\pi D(\tau - t)}} e^{-\frac{(x_f - z)^2}{4D(t_f - \tau)}} e^{-\frac{(z-x)^2}{4D(\tau - t)}} V(z)}{\dfrac{1}{\sqrt{4\pi D(t_f - t)}} e^{-\frac{(x_f - x)^2}{4D(t_f - t)}}}$$

$$= \int dz \frac{\sqrt{4\pi D(t_f - t)}}{\sqrt{4\pi D(t_f - \tau)}\sqrt{4\pi D(\tau - t)}} e^{-\frac{(x_f - z)^2}{4D(t_f - \tau)} - \frac{(z-x)^2}{4D(\tau - t)} + \frac{(x_f - x)^2}{4D(t_f - t)}} V(z)$$

$$=$$

and after some further simplifications, one obtains

$$\langle V(x(\tau))\rangle_0 = \int_{-\infty}^{+\infty} \frac{dz}{\sqrt{2\pi}} e^{-\frac{z^2}{2}} V(X + \alpha z) \tag{7.61}$$

where

$$X = \frac{x_f(\tau - t) + x(t_f - \tau)}{t_f - t}$$

$$\alpha = \sqrt{\frac{2D(t_f - \tau)(\tau - t)}{(t_f - t)}} \tag{7.62}$$

From Eq. (7.59), we have

$$\ln I(x) = \ln I_0(x) - D \int_t^{t_f} d\tau \langle V(x(\tau))\rangle_0 \tag{7.63}$$

and the Langevin bridge equation (7.39) becomes

$$\frac{dx}{dt} = \frac{x_f - x}{t_f - t} - 2D^2 \int_t^{t_f} d\tau \left(\frac{t_f - \tau}{t_f - t}\right) \int_{-\infty}^{+\infty} \frac{dz}{\sqrt{2\pi}} e^{-\frac{z^2}{2}} \frac{\partial}{\partial X} V(X + \alpha z) + \eta(t)$$

$$\tag{7.64}$$

This approximate bridge equation is valid for short simulation times. It is a stochastic integro-differential equation, and since it depends only on $x(t)$, it is Markovian. We now present a further approximation which is justified if the transition path time shows clear time scale separation $\tau_{TPT} \ll \tau_K$. Indeed, in Eq. (7.64), the variable z is a normal Gaussian variable and is thus of order 1. The variable α in (7.62) is of the order of the diffusion radius of the particle during time t_f. On the order hand, for any intermediate times t and τ, the variable X is of the order of magnitude of the distance $x_f - x_i$ between the initial and final states. So if the transition path is almost ballistic, the diffusion radius is much smaller than the distance between the initial and final states $\sqrt{\frac{2D(t_f-\tau)(\tau-t)}{(t_f-t)}} \ll$ $x_f - x_i$, and one may neglect the term αz in V in Eq. (7.64). The z integral then becomes trivial, and the above equation simplifies to

$$\frac{dx}{dt} = \frac{x_f - x}{t_f - t} - 2D^2 \int_t^{t_f} d\tau \left(\frac{t_f - \tau}{t_f - t}\right) \frac{\partial}{\partial X} V(X) + \eta(t) \tag{7.65}$$

Note that at low temperature, β large, the effective potential $V(X) = \left(\frac{\beta}{2}\frac{\partial U}{\partial X}\right)^2 - \frac{\beta}{2}\frac{\partial^2 U}{\partial x^2}$ can be approximated by

$$V(X) \approx \left(\frac{\beta}{2}\frac{\partial U}{\partial X}\right)^2 \tag{7.66}$$

by neglecting the second derivative term.

We illustrate this approximation on the case of the double-well potential $U(x) = \frac{1}{4}(x^2 - 1)^2$ using the simpler form of the effective potential from Eq. (7.66). The barrier height is $\beta U^{\ddagger} = 2.5$ and the average transition path time as given by Eq. (6.79) is $t_{TPT} \approx 2.2$. In Figure (7.10), we show 3 examples of paths generated exactly using Eq. (7.49) (in black) and by the approximate Eq. (7.65) (in red), for 3 different times t_f.

For short times (Figure a), $t_f < t_{TPT}$, the approximate trajectory is undistinguishable from the exact one. For times of the order of the transition path time (Figure b), $t_f \approx t_{TPT}$, the approximate trajectory is very close to the exact one, while for a longer time (Figure c), $t_f > t_{TPT}$, some discrepancies start to appear.

Let us mention that all this formalism can be generalized to the case of many degrees of freedom. Denoting by $\{x_i\}$ the set of variables, the simplified low temperature effective potential is

$$V(\{x_i\}) = \left(\frac{\beta}{2}\right)^2 \sum_k \left(\frac{\partial U(\{x_i\})}{\partial x_k}\right)^2 \tag{7.67}$$

and the bridge equation (7.65) becomes

$$\frac{dx_i}{dt} = \frac{x_i^{(f)} - x_i}{t_f - t} - D^2\beta^2 \int_t^{t_f} d\tau \left(\frac{t_f - \tau}{t_f - t}\right) \sum_k \frac{\partial U(\{X_l\})}{\partial X_k} \frac{\partial^2 U(\{X_l\})}{\partial X_i \partial X_k} + \eta_i(t) \tag{7.68}$$

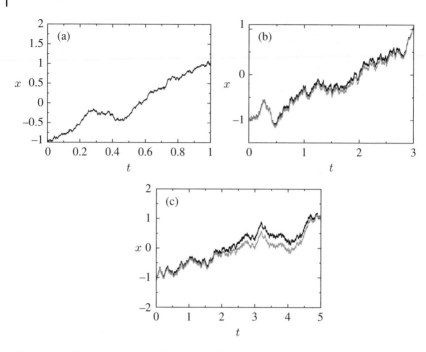

Figure 7.10 Three transition paths between the two minima of a double well potential, for short, average and long simulation times. Exact trajectories are in black, and approximate trajectories are in red. (*See color plate section for color representation of this figure*).

where

$$X_l = \frac{x_l^{(f)}(\tau - t) + x_l(t_f - \tau)}{t_f - t} \tag{7.69}$$

$$\langle \eta_l(t) \rangle = 0$$
$$\langle \eta_k(t)\eta_l(t') \rangle = 2D\delta_{kl}\delta(t - t') \tag{7.70}$$

This simplified equation requires the computation of the Hessian of the potential $U(\{x_i\})$.

To conclude this section, let us mention that these bridge equations are fairly simple to handle numerically. They provide statistically independent sets of transition paths. Other types of approximations than the ones considered here might be studied, and it is conceivable that these equations could be solved exactly for a large number of degrees of freedom.

References

1 Bender, C. and Orszag, S. (1978). *Advanced Mathematical Methods for Scientists and Engineers*. McGraw-Hill.

2 Zinn-Justin, J. (2002). *Quantum Field Theory and Critical Phenomena*. Clarendon Press.

3 Kirkpatrick, S., Gelatt, C.D., Vecchi, M.P. (1983). Optimization by simulated annealing, *Science* 220 (4598): 671–680.

4 Doob, J.L. (1957). *Bull. Soc. Math. France* 85: 431.

5 Fitzsimmons, P., Pitman, J., and Yor, M. (1993). Markovian bridges: construction, Palm interpretation, and splicing. In: *Seminar on Stochastic Processes 1992, Progress in Probability*, 33 101. Boston: Birkhäuser.

6 Majumdar, S.N. and Orland, H. (2015). Effective Langevin equations for constrained stochastic processes, *J. Stat. Mech: Theory Exp.* 2015.

7 Orland, H. (2011). Generating transition paths by Langevin bridges, *J. Chem. Phys.* 134: 174114.

8 Delarue, M., Koehl, P. and Orland, H. (2017). Ab initio sampling of transition paths by conditioned Langevin dynamics, *J. Chem. Phys.* 147: 152703.

Appendix A: Gaussian Variables

In probability theory, a Gaussian variable x is a random variable distributed according to a *normal distribution*

$$P(x) = \frac{1}{\sqrt{2\pi \ \sigma^2}} e^{-\frac{(x-\bar{x})^2}{2\sigma^2}} \tag{7.71}$$

This probability distribution is normalized and

$$\bar{x} = \langle x \rangle = \int_{-\infty}^{+\infty} dx \ \ xP(x) \tag{7.72}$$

$$\sigma^2 = \langle x^2 \rangle - \langle x \rangle^2 \tag{7.73}$$

An important equation that is used throughout the book is the *Gaussian integral*

$$\int_{-\infty}^{+\infty} \frac{dx}{\sqrt{2\pi \ \sigma^2}} e^{-\frac{(x-\bar{x})^2}{2\sigma^2}+ux} = e^{u\bar{x}+\sigma^2 \frac{u^2}{2}} \tag{7.74}$$

for any number u.

The normal distribution can be generalized to N Gaussian variables

$$P(x_1, \dots, x_N) = \left(\frac{1}{2\pi}\right)^{\frac{N}{2}} \frac{1}{\sqrt{\text{Det } C_{ij}}} e^{-\frac{1}{2}\sum_{ij}(x_i - \bar{x}_i)C_{ij}^{-1}(x_j - \bar{x}_j)} \tag{7.75}$$

where C_{ij} is a symmetric positive definite matrix, with inverse C_{ij}^{-1} and determinant Det C_{ij}. In addition, it is easily seen that

$$\bar{x}_i = \int dx_1 \dots dx_n \ x_i P(x_1, \dots, x_N) \tag{7.76}$$

$$C_{ij} = \langle(x_i - \bar{x}_i)(x_j - \bar{x}_j)\rangle - \langle(x_i - \bar{x}_i)\rangle\langle(x_j - \bar{x}_j)\rangle \tag{7.77}$$

and the matrix C_{ij} is called the *covariance matrix* of the variables $x_1, x_2, \dots, x_n$. If the matrix C_{ij} is not diagonal, the variables $x_1, \dots, x_n$ are said not to be independent.

Equation (7.74) can be easily generalized to

$$\int dx_1 \dots dx_n P(x_1, \dots, x_N)e^{\sum_i u_i x_i} = e^{\sum_i u_i \bar{x}_i + \frac{1}{2}u_i C_{ij} u_j} \tag{7.78}$$

Consider now a sum of N independent random variables $x_1, x_2, \dots, x_N$, where each x_i is Gaussian distributed with mean $\bar{x}_i$ and variance σ_i

$$P_i(x_i) = \sqrt{\frac{1}{2\pi\sigma_i^2}}e^{-\frac{(x_i - \bar{x}_i)^2}{2\sigma_i^2}} \tag{7.79}$$

Consider the random variable X, a linear combination of these random variables with coefficients λ_i

$$X = \sum_{i=1}^{N} \lambda_i x_i \tag{7.80}$$

The probability distribution of X can be obtained as

$$P(X) = \langle\delta(X - \sum_{i=1}^{N} \lambda_i x_i)\rangle \tag{7.81}$$

where the symbol $\langle\dots\rangle$ denotes the average over all Gaussian variables x_i. Using the Fourier representation of the δ-function

$$\delta(x) = \int_{-\infty}^{+\infty} \frac{dk}{2\pi}e^{ikx} \tag{7.82}$$

we have

$$P(X) = \int_{-\infty}^{+\infty} \frac{dk}{2\pi}\langle e^{ik(X - \sum_{i=1}^{N} \lambda_i x_i)}\rangle$$

$$= \int_{-\infty}^{+\infty} \frac{dk}{2\pi} \int \prod_i \frac{dx_i}{\sqrt{2\pi\sigma_i^2}}e^{-\frac{(x_i - \bar{x}_i)^2}{2\sigma_i^2}}e^{ik(X - \sum_{i=1}^{N}\lambda_i x_i)} \tag{7.83}$$

The Gaussian integrals can be performed using Eq. (7.78) and the result is

$$P(X) = \frac{1}{\sqrt{2\pi\sigma^2}} e^{-\frac{(X-\bar{X})^2}{2\sigma^2}} \tag{7.84}$$

where

$$\bar{X} = \sum_{i=1}^{N} \lambda_i \bar{x}_i = \langle X \rangle \tag{7.85}$$

$$\sigma^2 = \sum_{i=1}^{N} \lambda_i^2 \sigma_i^2 = \langle X^2 \rangle - \langle X \rangle^2 \tag{7.86}$$

Any linear combination of independent Gaussian variables is a Gaussian variable, the mean and variance of which are given by Eqs. (7.85) and (7.86). This theorem holds also if the Gaussian variables are not independent, as in Eq. (7.75). In that case, we find that $P(X)$ has the form of Eq. (7.84) with

$$\bar{X} = \sum_{i=1}^{N} \lambda_i \bar{x}_i = \langle X \rangle \tag{7.87}$$

$$\sigma^2 = \sum_{i,j} \lambda_i C_{ij} \lambda_j = \langle X^2 \rangle - \langle X \rangle^2 \tag{7.88}$$

Appendix B

In this appendix, we derive the notation

$$(dB(t))^2 = 2Ddt \tag{7.89}$$

useful to prove the Itô derivative formula.

For any function $f(x)$, define J as

$$J = \int_0^t (dB(s))^2 f(x(s)) \tag{7.90}$$

By discretization, we have

$$J = \sum_{n=0}^{N-1} (B(\tau_{n+1}) - B(\tau_n))^2 f(x(\tau_n)) \tag{7.91}$$

where $\tau_1, ..., \tau_N$ is a set of points of the interval $[0, t]$ with $\tau_n = ndt$ and $dt = t/N$. Note that since $(B(\tau_{n+1}) - B(\tau_n))^2 \sim dt$, we can use any value of s in the interval $[\tau_n, \tau_{n+1}]$ in $f(x(s))$ in eq. (7.90), the correction being of order $dt^{3/2}$.

Let us calculate the first two moments of the random variable J. We have

$$\bar{J} = \langle J \rangle = \sum_{n=0}^{N-1} \langle (B(\tau_{n+1}) - B(\tau_n))^2 f(x(\tau_n)) \rangle \tag{7.92}$$

Since the variable $x(\tau_n)$ does not depend on $B(\tau_{n+1}) - B(\tau_n)$, the expectation value factorizes and we have

$$\bar{J} = \sum_{n=0}^{N-1} \langle (B(\tau_{n+1}) - B(\tau_n))^2 \rangle \langle f(x(\tau_n)) \rangle$$

$$= 2Ddt \sum_{n=0}^{N-1} \langle f(x(\tau_n)) \rangle \tag{7.93}$$

where we have used

$$\langle (B(\tau_{n+1}) - B(\tau_n))^2 \rangle = 2Ddt \tag{7.94}$$

In the continuous limit, we have thus

$$\bar{J} = 2D \int_0^t ds\, f(x(s)) \tag{7.95}$$

Let us now compute the variance of the random variable J. We have (in discretized form)

$$\langle (J - \bar{J})^2 \rangle = \sum_{m=0}^{N-1} \sum_{n=0}^{N-1} \langle ((B(\tau_{m+1}) - B(\tau_m))^2 - 2Ddt)((B(\tau_{n+1}) - B(\tau_n))^2 - 2Ddt) f(x(\tau_m)) f(x(\tau_n)) \rangle$$

$$= \sum_{m=0}^{N-1} \sum_{n=0}^{N-1} \langle ((B(\tau_{m+1}) - B(\tau_m))^2 - 2Ddt)((B(\tau_{n+1}) - B(\tau_n))^2 - 2Ddt) \rangle \langle f(x(\tau_m)) f(x(\tau_n)) \rangle$$

where we have used again the factorization of the expectation values.

In the above sum, we have to distinguish the case $m = n$ from the case $m \neq n$

$$\langle (J - \bar{J})^2 \rangle = \sum_{m \neq n=0}^{N-1} \langle ((B(\tau_{m+1}) - B(\tau_m))^2 - 2Ddt) \rangle \langle ((B(\tau_{n+1}) - B(\tau_n))^2 - 2Ddt) \rangle \langle f(x(\tau_m)) f(x(\tau_n)) \rangle$$

$$+ \sum_{n=0}^{N-1} \langle ((B(\tau_{n+1}) - B(\tau_n))^2 - 2Ddt)^2 \rangle \langle (f(x(\tau_n)))^2 \rangle$$

The first term above vanishes and due to Eq. (7.94), we have

$$\langle (J - \bar{J})^2 \rangle = \sum_{n=0}^{N-1} \langle ((B(\tau_{n+1}) - B(\tau_n))^2 - 2Ddt)^2 \rangle \langle (f(x(\tau_n)))^2 \rangle$$

We have

$$\langle ((B(\tau_{n+1}) - B(\tau_n))^2 - 2Ddt)^2 \rangle = \langle (B(\tau_{n+1}) - B(\tau_n))^4 \rangle$$
$$- 4Ddt \langle (B(\tau_{n+1}) - B(\tau_n))^2 \rangle + 4D^2 dt^2$$

Since the random noise $B(\tau_{n+1}) - B(\tau_n)$ is a Gaussian variable, we have

$$\langle (B(\tau_{n+1}) - B(\tau_n))^4 \rangle = 3 \langle (B(\tau_{n+1}) - B(\tau_n))^2 \rangle^2$$
$$= 3(2Ddt)^2 = 12D^2 dt^2$$

so that finally

$$\langle (J - \bar{J})^2 \rangle = 8D^2 dt^2 \sum_{n=0}^{N-1} \langle (f(x(\tau_n)))^2 \rangle$$

In the continuous limit, we obtain

$$\langle (J - \bar{J})^2 \rangle = 8D^2 dt \int_0^t d\tau \langle (f(x(\tau)))^2 \rangle$$

This result shows that the variance of the random variable J goes to 0 as $dt \to 0$. Therefore, in the continuous limit, J is constant (not random), $J = \bar{J}$ and for any regular enough function f, we have

$$\int_0^t (dB(s))^2 f(x(s)) = 2D \int_0^t ds \, f(x(s)) \tag{7.96}$$

Taking a derivative of this equation with respect to t, we see that this identity can be symbolically written as

$$(dB(t))^2 = 2D dt \tag{7.97}$$

in any stochastic integral.

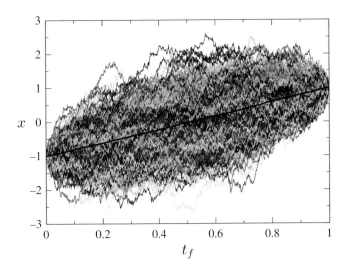

Figure 7.6 Brownian bridges.

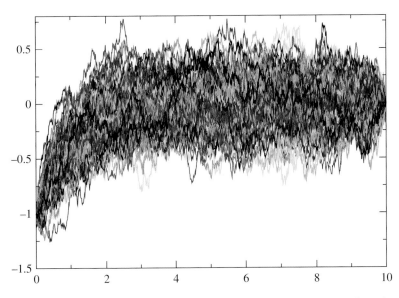

Figure 7.8 A sample of 100 Ornstein-Uhlenbeck trajectories between -1 and 0 with $t_f = 10$.

Molecular Kinetics in Condensed Phases: Theory, Simulation, and Analysis,
First Edition. Ron Elber, Dmitrii E. Makarov and Henri Orland.
© 2020 John Wiley & Sons Ltd. Published 2020 by John Wiley & Sons Ltd.

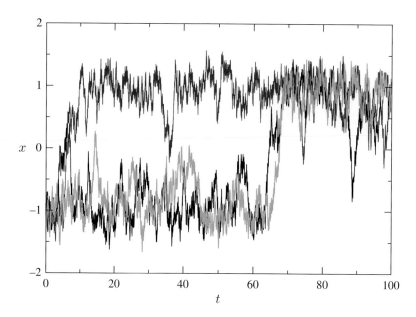

Figure 7.9 Three bridge trajectories in a double well, joining the two local minima.

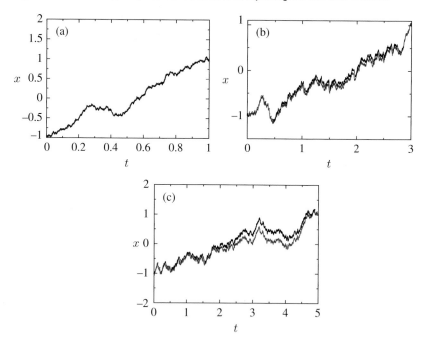

Figure 7.10 Three transition paths between the two minima of a double well potential, for short, average and long simulation times. Exact trajectories are in black, and approximate trajectories are in red.

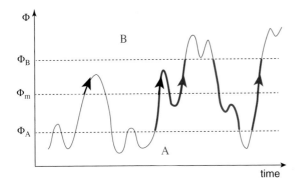

Figure 8.3 Mapping continuous trajectory onto two-state dynamics. Red and blue trajectory pieces are assigned as A and B states. Transition paths are shown in bold. Arrows indicate when the trajectory crosses the mid-point Φ_m in the direction from A to B. All these crossings contribute to the transition-state-theory rate, Eq. (8.15), but only the crossings indicated by the red arrows will contribute to Eq. (8.17).

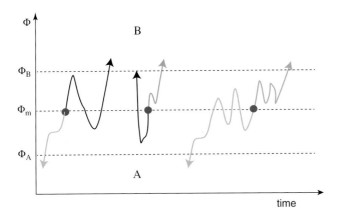

Figure 8.4 Generation of transition paths by shooting forward and backward from intermediate points located on the hypersurface $\Phi(\mathbf{x}) = \Phi_m$. In transition state theory, every trajectory with a positive trajectory is a transition path. Here, only the rightmost trajectory is accepted; the two rejected trajectories contain invalid segments shown in black.

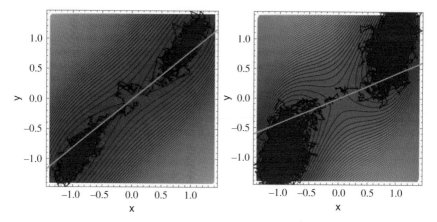

Figure 11.3 Numerically simulated trajectories undergoing Brownian dynamics in a potential of Eq. (11.1) (with $U(x)$ being a symmetric double well, which is the same in both cases). The spring constant K is lower in the case depicted on the right and the friction coefficients are the same for x and y, $\gamma_x = \gamma_y$. The green line shows the direction of the unstable mode $\mathbf{e}_1$.

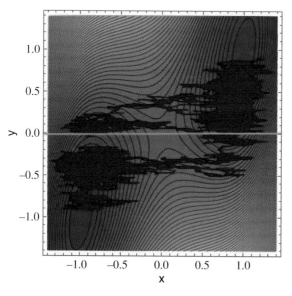

Figure 11.4 Numerically simulated trajectories undergoing Brownian dynamics for the same potential as in the right panel of Figure 11.3, but with the friction coefficient along y increased 50-fold, i.e., $\gamma_y = 50\gamma_x$. The green line shows the direction of the unstable mode $\mathbf{e}_1$.

Figure 16.1 A two-dimensional contour plot of the Mueller energy function. [1] The contour lines are of equi-potential values, and the color code varies from blue to yellow, where the blue is the lowest energy. The yellow domain is of high energy and is inaccessible to trajectories at a moderate temperature.

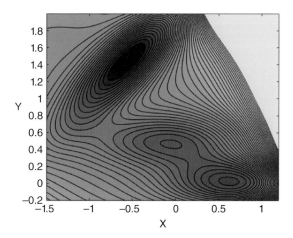

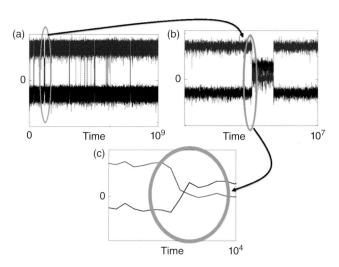

Figure 16.3 Transitions between states in the Mueller potential extracted from a long trajectory are displayed at different magnifications of the time scale. The black and red lines are the trajectory values along the horizontal and vertical axes as a function of time respectively. The lowest energy minimum is populated when the black and red lines are well separated and the black curve is negative. The time window is magnified as we shift from panels A to C. Note the difference in time scales and the blue circles around a transition that we choose to magnify. Similar pictures illustrating metastability can be found elsewhere in the book, see Figure 1.2 and 8.2.

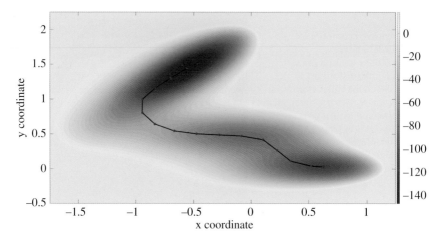

Figure 17.1 A minimum energy path computed by the locally updated planes approach described in reference [3] and Eq. 17.3 from the top left minimum of the Mueller potential to the lower right minimum. The reaction coordinate is presented by a set of discrete configurations along the curve (blue dots). The connecting dashed line is to guide the eye. The colors code the energy value with the red the lowest energies. An energy scale is provided at the right bar.

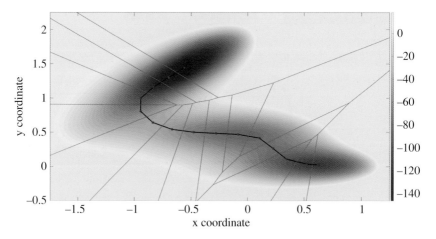

Figure 18.1 The Mueller potential and its partition to Voronoi cells using configurations along a reaction coordinate. The black line is a discrete reaction coordinate computed with Eq. (17.3). The blue dots along the line are the discrete optimized points that are also used as centers of Voronoi cells. The green straight-line segments are the boundaries of the Voronoi cells that are also used as milestones.

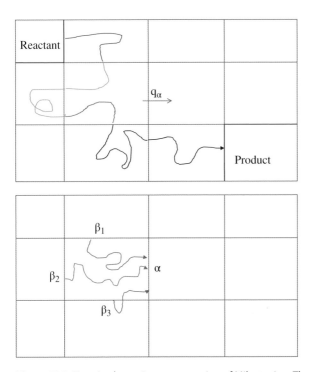

Figure 18.2 Top: A schematic representation of Milestoning. The space is partitioned to cells and trajectories are conducted between boundaries of cells. The quantity of interest is the flux, q_α, which is the number of trajectories that cross milestone α in unit time. It is also possible to consider a single long reactive trajectory and to chop it into pieces between boundaries as illustrated for the colored trajectory fragments. Bottom: A schematic representation of Milestoning trajectories that are used to estimate the flux in Eq. (18.1). Trajectories are initiated at milestones $\beta_1, \beta_2, \beta_3$ and their time courses are followed until they hit for the first time milestone α. Note that the trajectories are allowed to cross the milestone they were initiated at. The total number of trajectories that cross milestone α at time τ is the summation over all the trajectories that arrived to it for the first time at τ. Using the trajectories, we can estimate the kernel $w_{\beta, \alpha}(\mathbf{y}, t; \mathbf{x}, \tau)$.

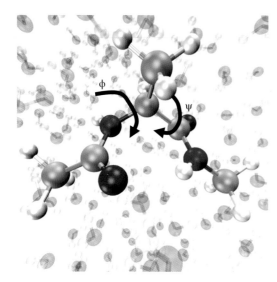

Figure 19.1 A space-filling model of an alanine dipeptide molecule embedded in a water box. The transparent pink spheres are the oxygen atoms of the water molecules. The big red and solid spheres are carbonyl oxygens and the blue spheres are nitrogen atoms. The rotation around bonds, ϕ and ψ are illustrated with the curved arrows and are the coarse variables that are typically used in studies of conformational transitions in protein and peptide backbones. The figure was prepared with the software VMD [1].

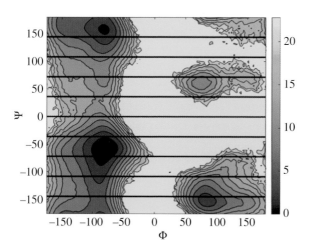

Figure 19.2 The free energy landscape of a solvated alanine dipeptide is shown as a function of the two coarse variables, ϕ and ψ. Note the existence of two deep minima on the left side of the map. The lower minimum on the left (negative ψ) corresponds to an α helix, a common secondary structure element of proteins, while the upper minimum on the left (positive ψ) is of an extended chain conformation (or a secondary structure of β sheet). The free energy depends only weakly on ϕ. Therefore, we reduce the number of coarse variables to one, ψ. Milestones along the ψ dihedral angle are shown as black lines parallel to the ϕ axis.

8

The Rate of Conformational Change: Definition and Computation

8.1 First-order Chemical Kinetics

Chemistry studies how molecules interconvert between their different forms. The description of this process usually takes the form of phenomenological rate equations describing the time evolution of concentrations (or populations) of chemical species, which is determined by rate coefficients (also referred to as "rate constants" or even simply "rates"). Such coefficients have already been introduced here: Section 4.1 introduced rate matrices, whose elements describe interconversion rates between pairs of discrete states, and Section 6.2 discussed the escape rate from a metastable state. Because the concept of rate is cental to chemical kinetics, this Chapter discusses how to define the rate precisely, how to relate the rate to the underlying microscopic dynamics, and how to compute the rate efficiently, both exactly and with simple approximations.

Consider the simple example of chair-boat isomerization (Figure. 8.1), where the chemical identity of the molecule remains unchanged, but the geometry is altered. Chemists describe this reaction by the following scheme:

$$A \underset{k_{B \to A}}{\overset{k_{A \to B}}{\rightleftarrows}} B \tag{8.1}$$

Many other examples described by Eq. (8.1) exist, both in chemistry and outside it. Importantly, unless A and B denote specific quantum states of the same molecule, they do not provide precise information about its state. In some cases (such as in Figure. 8.1), the molecular forms A and B describe specific molecular geometries, to within relatively small fluctuations caused by thermal motion or quantum effects. But in other cases, A and B may represent ensembles involving diverse structures. For example, when the scheme of Eq. (8.1) is used to describe protein folding, either A or B represents an unfolded protein, a molecular chain that can assume an astronomical number of random conformations.

Usually molecular species A and B are only well defined when they remain stable over a sufficiently long time, during which they can be detected experimentally. This time is long enough for the molecule to achieve partial thermal

Molecular Kinetics in Condensed Phases: Theory, Simulation, and Analysis,
First Edition. Ron Elber, Dmitrii E. Makarov and Henri Orland.
© 2020 John Wiley & Sons Ltd. Published 2020 by John Wiley & Sons Ltd.

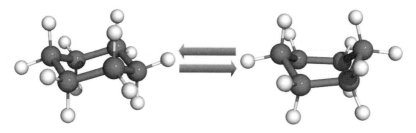

Figure 8.1 Chair-boat isomerization as an example of a reversible unimolecular reaction.

equilibrium within A or B. For example, it is reasonable to assume that thermal fluctuations of the molecule around its chair or boat structure in Figure 8.1 obey Boltzmann statistics. Let τ_r be a characteristic timescale of the molecule's relaxation within A or B and suppose it is much shorter than the average dwell time $\langle t_{A(B)} \rangle$ within A(B). The time τ_r also sets a timescale over which the molecule loses memory of its past. Imagine following a long molecular trajectory that shows transitions between A and B starting from the instant $t = 0$ that the molecule has just arrived in A (Figure 8.2). After a time $\sim \tau_r$ the memory of this event is lost; the probability to undergo a transition to B should then be independent of the molecule's history. Let $k_{A \to B}dt$ be the (conditional) probability to make a transition between t and $t + dt$ provided that the molecule is found in A at time t. Provided that external conditions (such as temperature) are not changing, the coefficient $k_{A \to B}$ has no explicit time dependence. Similarly, we can introduce a coefficient $k_{B \to A}$, which is probability, per unit time, that a molecule with initial conditions randomly drawn from the Boltzmann ensemble within B will jump into A. Now let $P_A(t)$ and $P_B(t)$ be the probabilities to find the molecule in A and B. These probabilities depend on the initial conditions; for the specific initial condition where the molecule starts in A we have $P_A(0) = 1$; $P_B(0) = 0$, but as the time evolves we have less certainty about the molecule's state as a result of its transitions between the states. The probability of undergoing a jump from A to B after an infinitesimal time dt is $k_{A \to B}dt$ times the probability $P_A(t)$ to be in A at time t. Likewise, the probability of making a jump from B to A is $k_{B \to A}dtP_B(t)$. The change in the probability $P_A(t)$ as a result of such jumps is $P_A(t + dt) - P_A(t) = -k_{A \to B}dtP_A + k_{B \to A}dtP_B$, and a similar equation can be written for P_B. We conclude that these probabilities obey simple gain-loss equations

$$\frac{dP_A}{dt} = -k_{A \to B}P_A + k_{B \to A}P_B$$

$$\frac{dP_B}{dt} = k_{A \to B}P_A - k_{B \to A}P_B \tag{8.2}$$

We emphasize that these equations are coarse-grained both in time (in that we are considering evolution at times longer than τ_r) and in space (in that, all points

Figure 8.2 Mapping of a continuous molecular trajectory onto a two-state process.

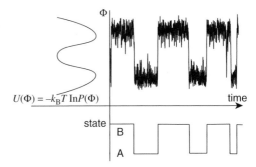

in the phase space of the molecule are lumped into only two possible states A or B). If we have a collection of $N \gg 1$ molecules that can be viewed as evolving independently each according to Eq. (8.2), as, for example, in a sufficiently dilute solution, then, by multiplying Eq. (8.2) by N and recognizing that $N_{A(B)} = NP_{A(B)}$ is the expectation value of the molecules of each type, we obtain the equations

$$\frac{dN_A}{dt} = -k_{A \to B}N_A + k_{B \to A}N_B$$

$$\frac{dP_B}{dt} = k_{A \to B}N_A - k_{B \to A}N_B. \tag{8.3}$$

If these equations are divided by the volume V of the system, one obtains similar equations for the concentrations, $[A(B)] = N_{A(B)}/V$, of each species, which is the form most familiar to chemists. These equations are known as the 1st order kinetics equations. The coefficients $k_{A \to B}$ and $k_{B \to A}$ are known as the rate coefficients (or, in other literature, "rate constants"). Predicting or measuring rate coefficients is the central goal of chemical kinetics.

8.2 Rate Coefficients from Microscopic Dynamics

By verifying that a mixture of chemicals A and B evolves according to Eq. (8.3), the experimentalist can measure the rate coefficients $k_{A \to B}$ and $k_{B \to A}$. A computational chemist, in contrast, can obtain molecular trajectories by solving the appropriate equations of motion. It is therefore necessary to find a mapping between the phenomenological picture of jumps between discrete states, as described by Eqs. (8.2–8.3), and the continuous molecular trajectories. More specifically, we need to address the following questions:

1. Although it is always possible to group all molecular states into two classes, A and B, this does not guarantee the validity of Eq. (8.2). Under which conditions does first order kinetics hold?
2. Given a molecular trajectory, how can the rate coefficients be computed from it in principle?

3. As it will transpire from the subsequent discussion, obtaining rate coefficients is a computationally demanding task. This is because conformational changes are often infrequent events, and a very long computer simulation may be required to observe even a single transition. How can the rate coefficients be computed efficiently?

4. Are there useful approximations relating rate coefficients to salient features of molecular energy landscapes?

We proceed by discussing each of these issues.

8.2.1 Validity of First Order Kinetics

An experimentalist studying chemical kinetics monitors a specific molecular property Φ of the molecule. This could, for example, be the intensity of the fluorescence light emitted by the molecule or some structural characteristic of the molecule. This property is a function of the molecular configuration, which we will denote by the vector $\mathbf{x}$. As the molecular configuration evolves in time, $\Phi = \Phi[\mathbf{x}(t)]$ is time dependent. Distinct molecular states are manifested by peaks in the probability distribution of Φ, or, equivalently, as minima in the potential of mean force $U(\Phi) = -k_B T \ln P(\Phi)$. We note that the quantity Φ is often referred as a "reaction coordinate"; how to define and compute "good" reaction coordinates is discussed later in this book (Chapter 17); experimentally, however, the choice of the reaction coordinate is often dictated by the measurement setup.

Figure 8.2 illustrates the dynamics as manifested by the time dependence of Φ. The trajectory shown in this figure was generated by assuming that Φ obeyed a Langevin equation from Chapter 1,

$$m\ddot{\Phi} = -U'(\Phi) - \gamma\dot{\Phi} + \zeta(t), \tag{8.4}$$

in a potential $U(\Phi)$ that has two wells separated by a barrier, whose height significantly exceeds the thermal energy $k_B T$. Because most of the time the system is found close to the minimum of one well (where the probability distribution of Φ is peaked), we can classify being in one well as state A and in the other as state B. The precise mathematical mapping between Φ and discrete states will be the subject of further discussion, but, at least in the situation depicted in Figure 8.2, simple visual inspection of the trajectory $\Phi(t)$ seems to provide a nearly unambiguous reduction to two states.

Consider a continuous piece of the trajectory contained within one of the wells, say A. As discussed before, there is a characteristic relaxation timescale τ_r over which the memory of the initial state is lost, and the molecule can be assumed to achieve thermal equilibrium within A. If this timescale is much shorter than the average dwell time within A, $\langle t_A \rangle$, then, for most of its sojourn in A, the molecule can be thought of as being in partial thermal equilibrium within A. Specifically, if the probability distribution $P(\Phi)$ is calculated only for this piece of trajectory, Boltzmann distribution will be observed, but only with a single peak corresponding to A and missing the peak corresponding from B.

If we are only interested in long timescales comparable with $\langle t_A \rangle$, we can coarse grain time and declare that we are unconcerned about what happens to the molecule at timescales comparable to τ_r or shorter. With this coarse view of time it no longer makes sense to talk about the instant molecular configuration or the precise value of Φ. Instead we think of a molecular configuration as randomly drawn from the (partial) equilibrium ensemble. We then ask: what is the probability that a molecule, with its state randomly drawn from this ensemble, will make a transition to B during a short time interval $\Delta t(\Delta t \ll \langle t_A \rangle)$? We expect this probability to be proportional to Δt, and the proportionality coefficient is the rate coefficient $k_{A \rightarrow B}$.

Under the time-scale separation assumption described above, we now formulate several equivalent definitions of the rate coefficients, which will be useful for practical rate calculations.

First, we ask: what is the probability density $P_{dwell}(t_A)$ of the dwell time t_A within A? Let $P_s(t)$ be the survival probability in state A, i.e. the probability that the molecule has not escaped A during a time interval t, provided it is in A at $t=0$. Then the probability of escaping A in an infinitesimal time interval between t_A and $t_A + dt_A$ is

$$P_{dwell}(t_A)dt_A = k_{A \rightarrow B}P_s(t_A)dt_A \tag{8.5}$$

On the other hand, this should be equal to the decrease in the survival probability, $P_s(t_A) - P_s(t_A + dt_A) = -dP_s$, resulting in a simple differential equation governing the survival probability, $dP_s/dt_A = -k_{A \rightarrow B}P_s$, which, together with the initial condition $P_s(0) = 1$, yields

$$P_s(t_A) = e^{-k_{A \rightarrow B}t_A},$$
$$P_{dwell}(t_A) = k_{A \rightarrow B}e^{-k_{A \rightarrow B}t_A} \tag{8.6}$$

The distribution of the dwell times within A (or, similarly, B) is thus exponential, with its average given by

$$\langle t_A \rangle = \int_0^\infty dt_A t_A P_{dwell}(t_A) = \frac{1}{k_{A \rightarrow B}}. \tag{8.7}$$

Similarly, we have $\langle t_B \rangle = 1/k_{B \rightarrow A}$ for the mean dwell time in the state B.

Second, consider the thought experiment in which the initial configuration of the molecule is drawn from the Boltzmann distribution within A. We now ask: what is the mean of the time to escape from state A and arrive in B? This quantity is called the mean first passage time or (MFPT) from A to B (cf. Section 6.1). The answer is still given by Eq. (8.7) because we are dealing with a memoryless process, and so the past of the molecule prior to this experiment is irrelevant. The equivalence of the mean dwell time and the mean first passage time hinges upon the assumed fast loss of memory after the molecule first arrives in A – the memory of this transition is lost and the molecule can be assumed to be in the generic state A.

Third, consider now a long trajectory of the molecule which contains many jumps from A to B and from B to A. Let $n_{A \to B}(t)$ ($n_{B \to A}(t)$) be the number of A-to-B (B-to-A) transitions observed over a long period of time t, $n_{A \to B}(t)$. The ratio $q_{A \to B} = n_{A \to B}(t)/t$, equal to the number of transitions per unit time, can be interpreted as a unidirectional flux from A to B. Assuming the time reversal symmetry of the underlying dynamics (see next Chapter for more on this topic), one immediately arrives at an important result:

$$q_{A \to B} = q_{B \to A} \tag{8.8}$$

To estimate the flux $q_{A \to B}$, we envision $N \gg 1$ replicas of the same system and ask how many of them will jump from A to B during a time interval Δt. The answer is $N_A k_{A \to B} \Delta t$, where the number of molecules in A is $N_A = N P_A^{eq}$ and

$$P_A^{eq} = \frac{\langle t_A \rangle}{\langle t_A \rangle + \langle t_B \rangle} = \frac{k_{A \to B}^{-1}}{k_{A \to B}^{-1} + k_{B \to A}^{-1}},$$

P_A^{eq} being the equilibrium probability of being in A (equal to the fraction of time spend in A). The flux, being the number of jumps per unit time per molecule is then $k_{A \to B} P_A^{eq}$, or

$$k_{A \to B} = q_{A \to B}/P_A^{eq} \tag{8.9}$$

Summarizing the above arguments, we can propose several equivalent ways to calculate $k_{A \to B}$:

1. Start in A with a canonical distribution and calculate the mean first passage time to arrive in B, $MFPT(A \to B)$, by averaging over initial conditions. We obtain $k_{A \to B} = 1/MFPT(A \to B)$.
2. Monitor a single trajectory over a long time, record dwell times within A, and calculate the average. We find $k_{A \to B} = 1/\langle t_A \rangle$.
3. For a single trajectory, calculate the flux (the number of transitions per unit time) from A to B and divide it by the equilibrium population of A.

The astute reader will notice that the rate definitions 1 and 3 have already been used in the calculation of the escape rate for the Kramers problem in Chapter 6. Under the time-separation assumption above, all three definitions are equivalent. Each definition has its advantages, depending on whether we wish to estimate the rate coefficients numerically or analytically and on the specifics of model used to describe the dynamics of the system.

8.2.2 Mapping Continuous Trajectories onto Discrete Kinetics and Computing Exact Rates

We now tackle the task of mapping a continuous trajectory $\Phi(t)$ onto a stochastic process involving just two states, A and B (Figure 8.2). The simplest (and naïve) solution is to pick some intermediate value Φ_m, and to assign the values $\Phi < \Phi_m$ and $\Phi > \Phi_m$, respectively, to A and B. Using the definition of the rate given by Eq. (8.9), we can estimate the transition rate if we know the number of times the line $\Phi = \Phi_m$ is crossed, per unit time, with a positive velocity, $\dot{\Phi} > 0$, corresponding to transitions from A to B. Suppose, for example, that the dynamics along Φ is governed by the Langevin equation, Eq. (8.4). An essential feature of this equation is that the resulting equilibrium distribution of the coordinate Φ and the velocity $\dot{\Phi}$ is the Maxwell-Boltzmann distribution corresponding to the potential $U(\Phi)$. To evaluate $q_{A \to B}$, we envision, again, $N \gg 1$ replicas of the same system, each evolving stochastically under Eq. (8.4). How many of these replicas are crossing the boundary Φ_m during a short time interval Δt? To cross this boundary, the molecule has to be within the distance $\dot{\Phi}\Delta t = p_\Phi \Delta t/m$ from the boundary, where p_Φ is the momentum associated with Φ. Here, we assume the time interval Δt to be short enough to neglect the change in momentum as the system travels this distance. The number of such molecules is given by a Boltzmann average over the momentum

$$\Delta N = \frac{N}{Z} \int_0^\infty \frac{dp_\Phi}{2\pi\hbar} e^{-\beta \frac{p_\Phi^2}{2m} - \beta U(\Phi_m)} \frac{p_\Phi}{m} \Delta t = \frac{N\Delta t}{Z} \frac{1}{2\pi\hbar\beta} e^{-\beta U(\Phi_m)}, \tag{8.10}$$

where

$$Z = \int_{-\infty}^{\infty} d\Phi \int_0^\infty \frac{dp_\Phi}{2\pi\hbar} e^{-\beta \frac{p_\Phi^2}{2m} - \beta U(\Phi_m)} \tag{8.11}$$

is the partition function of the system and $\beta = (k_B T)^{-1}$ is the inverse thermal energy. Thus, we have

$$q_{A \to B} = \frac{1}{Z} \frac{1}{2\pi\hbar\beta} e^{-\beta U(\Phi_m)} \tag{8.12}$$

To use Eq. (8.9), we note that the equilibrium probability to be in A can be written as

$$P_A^{eq} = \frac{Z_A}{Z} = \frac{\int_{-\infty}^{\Phi_m} e^{-\beta U(\Phi)} d\Phi}{\int_{-\infty}^{+\infty} e^{-\beta U(\Phi)} d\Phi}, \tag{8.13}$$

where Z_A is the partition function of molecule A (in which the integration over Φ is restricted to the space to the left of Φ_m). Then, using Eqs. (8.9 and 8.12) we obtain

$$k_{A \to B} = \frac{k_B T}{2\pi\hbar Z_A} e^{-\beta U(\Phi_m)} \tag{8.14}$$

In a more general case of arbitrary dynamics along Φ (as long as the dynamics is consistent with the Maxwell-Boltzmann statistics) we can generalize this equation by writing it as a Boltzmann average

$$k_{A \to B} = \frac{1}{P_A^{eq}} \langle \dot{\Phi} \theta(\dot{\Phi}) \delta(\Phi - \Phi_m) \rangle = \frac{Z}{Z_A} \left\langle \nabla \Phi \frac{\mathbf{p}}{m} \theta \left[\nabla \Phi \frac{\mathbf{p}}{m} \right] \delta(\Phi - \Phi_m) \right\rangle.$$

(8.15)

Here $\Phi = \Phi(\mathbf{x})$ is viewed as some specified function of the molecule's configuration vector $\mathbf{x}$, the velocity of the system is written as $\dot{\mathbf{x}} = \mathbf{p}/m$, and the Heaviside theta function θ restricts the phase-space integration to phase-space points with positive velocities $\dot{\Phi}$. The angular brackets denote the Boltzmann average,

$$\langle \dots \rangle = Z^{-1} \int \frac{d^N \mathbf{x} d^N \mathbf{p}}{(2\pi\hbar)^N} e^{-\beta H(\mathbf{x}, \mathbf{p})} (\dots),$$

(8.16)

where N is the dimensionality of the system, and where equilibrium distribution under a Hamiltonian $H(\mathbf{x}, \mathbf{p})$ is assumed.

A particularly useful simplification of Eq. (8.15) is achieved when $\Phi(\mathbf{x})$ is a linear function of the atomic Cartesian coordinates, $\Phi(\mathbf{x}) = \mathbf{e}\mathbf{x} - c$, so that $\nabla \Phi = \mathbf{e}$ is constant. In this case the surface $\Phi = \Phi_m$ separating the reactants from products is a plane orthogonal to the vector $\mathbf{e}$. Because, in the case of the Maxwell-Boltzmann statistics, the distributions of the atomic coordinates and momenta are statistically independent, we can rewrite Eq. (8.15) as a product of two averages, one over the momenta and the other over the coordinates:

$$k_{A \to B} = \frac{1}{P_A^{eq}} \langle \dot{\Phi} \theta(\dot{\Phi}) \rangle \rangle \langle \delta(\Phi - \Phi_m) \rangle = \frac{1}{2 P_A^{eq}} \langle |\dot{\Phi}| \rangle P(\Phi_m)$$

(8.15a)

Here $P(\Phi_m) = \langle \delta(\Phi - \Phi_m) \rangle$ is the normalized equilibrium distribution of Φ and $\langle |\dot{\Phi}| \rangle$ is the average absolute value of the thermal velocity measured along Φ. In particular, using the equation $\langle |p_\Phi| \rangle = \sqrt{2 k_B T m / \pi}$, our result for the one-dimensional model of a particle of mass m moving along Φ, Eq. (8.14), can be recast in the form:

$$k_{A \to B} = \frac{1}{2 P_A^{eq}} \sqrt{\frac{2 k_B T}{\pi m}} P(\Phi_m)$$

(8.14a)

Equations (8.14, 8.14a and 8.15, 8.15a) express what is known as the *transition state theory* (TST)[1–3], which will be discussed in more detail below. Further examination of these equations, however, reveals several problems. First, the rate coefficient predicted by these equations depends on the selection of the point Φ_m partitioning the space into A and B, contrary to the expectation that $k_{A \to B}$ should be independent of the precise way it is measured and of the precise way how A and B are defined. Second, Eqs. (8.14–8.15) are entirely determined by the equilibrium statistical-mechanical properties

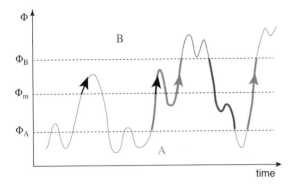

Figure 8.3 Mapping continuous trajectory onto two-state dynamics. Red and blue trajectory pieces are assigned as A and B states. Transition paths are shown in bold. Arrows indicate when the trajectory crosses the mid-point Φ_m in the direction from A to B. All these crossings contribute to the transition-state-theory rate, Eq. (8.15), but only the crossings indicated by the red arrows will contribute to Eq. (8.17). (*See color plate section for color representation of this figure*).

of the molecules and are independent of any parameters characterizing the dynamics. For the Langevin case, Eq. (8.4), in particular, these equations imply independence of $k_{A \rightarrow B}$ on the friction coefficient γ. This does not make sense: imagine that the reaction in question takes place in a viscous solution. Then simply increasing the solvent viscosity (and thus increasing the friction coefficient) is expected to slow the reaction down (Chapter 6), an effect that our theory fails to predict. What is wrong?

The problem is that we did not count the transitions properly. Indeed, Figure 8.3 shows that counting crossings of some intermediate point between A and B regions fails to properly identify transitions in two ways. First, the trajectory may cross Φ_m in the direction from A to B, but then quickly cross it back, having failed to reach B. By reaching B we mean settling within the B basin of attraction for a time significantly longer than τ_r, as our physical interpretation of the rate coefficients requires. Our theory, however, will count the first of the two crossings observed in Figure 8.3 as true transitions. Second, a trajectory undergoing a single transition from A to B may recross Φ_m multiple times. If such rapid recrossings take place on a timescale comparable to or shorter than τ_r, then the coarse graining arguments presented above require that we lump them into a single transition, while our approach so far has counted them as multiple transitions.

From a number of equivalent ways to fix the problem of overcounting crossings [4–6], here we will describe one method that translates into an efficient computational scheme, allows us to make certain exact statements about the transition state theory approximation, and is closely related to other methods described later in this book.

First, we define the transition region between A and B as all conformations satisfying the relationship $\Phi_A < \Phi < \Phi_B$. The precise values of Φ_A and Φ_B are not important, but it is assumed that a trajectory undergoing a transition and crossing one of these boundaries, say A, is committed to the respective state (A) in that a time of order of $1/k_{A \to B}$ is required before the next transition from A to B will take place. We now define a *transition path* as a piece of a trajectory that enters the transition region from A (B) and exits it through B (A) without exiting the transition region in between these two events. Such transition paths have already been discussed in Chapters 6 and 7. The trajectory shown in Figure 8.3 includes three transition paths shown in bold.

To obtain the "exact" transition rate coefficient, we identify the fluxes $q_{A \to B}$ ($q_{B \to A}$) as the number of transition paths from A to B (B to A) observed per unit time. When, for example, the leftmost transition path in Figure 8.3 crosses the line Φ_m 3 times, it is only counted once. Similarly, a crossing of the transition region boundaries or of the point Φ_m is not counted if it is not part of a transition path.

A convenient way to think of the mapping between the continuous function $\Phi(t)$ and a two-state process that corresponds to the above interpretation of the fluxes is as follows: For any point on the trajectory we assign the corresponding state (A or B) such that the state switches to, say, A when the boundary Φ_A is crossed, in the direction from B to A, *for the first time* after crossing the other boundary. Subsequent recrossings of this boundary do not change the state assignment until the boundary Φ_B is crossed for the first time, and then the state assignment is changed to B. It is easy to see that each such state switch is preceded by a transition path, and so this state assignment, indeed, corresponds to our definition of the rate.

8.2.3 Computing the Rate More Efficiently

The new definition of the rate coefficient $k_{A \to B}$ does not, in principle, involve the mid-point Φ_m. Counting transitions as crossings of this point is, however, advantageous for numerical reasons. To see this, consider first 8.15: it expresses a statistical mechanical average, evaluation of which does not require computation of the time evolution of the system $\mathbf{x}(t)$. One can, for example, evaluate it using a Monte Carlo procedure: Sample the momentum $\mathbf{p}$ and the position $\mathbf{x}$ from the Boltzmann distribution subject to the hypersurface constraint $\Phi(\mathbf{x}) = \Phi_m$ and perform a Monte Carlo average of $\nabla\Phi\frac{\mathbf{p}}{m}\theta\left[\nabla\Phi\frac{\mathbf{p}}{m}\right]$ over all such trajectories; the Heaviside function ensures that only the trajectories with positive values of the initial velocity, $\dot{\Phi} = \nabla\Phi\mathbf{p}/m$, will contribute into this average.

The above procedure will effectively sample all positive-velocity crossings of Φ_m, as indicated by arrows in Figure 8.3. To obtain the correct rate $k_{A \to B}$, however, only the last positive-velocity crossings of transition paths (shown as red arrows in Figure 8.3) must be sampled. To accomplish this, we modify the

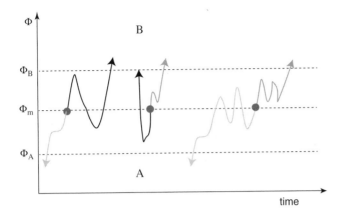

Figure 8.4 Generation of transition paths by shooting forward and backward from intermediate points located on the hypersurface $\Phi(\mathbf{x}) = \Phi_m$. In transition state theory, every trajectory with a positive trajectory is a transition path. Here, only the rightmost trajectory is accepted; the two rejected trajectories contain invalid segments shown in black. (*See color plate section for color representation of this figure*).

sampling procedure as follows: Randomly sample (with Boltzmann weights) phase-space points $\mathbf{x}(0)$, $\dot{\mathbf{x}}(0) = \mathbf{p}(0)/m$ satisfying the constraint $\Phi(\mathbf{x}) = \Phi_m$ and use them as initial conditions to launch trajectories forward and backward in time (with the reversed initial momenta - $\mathbf{p}(0)$ used to launch backward trajectories). Taking advantage of the time reversal symmetry, an A-to-B transition path can be constructed from a forward-propagated piece that must end on the boundary Φ_B and a time-reversed, backward-propagated piece originating with a momentum - $\mathbf{p}(0)$ (Figure 8.4). Moreover, the time $t = 0$ in this calculation must correspond to the last crossing of Φ_m. For the reason, the forward–time trajectory can be propagated until it either recrosses Φ_m (in which case it is rejected) or arrives at the boundary Φ_B; in the latter case we still need to check whether the time-reversed trajectory satisfies the requirement that it reaches Φ_A before escaping the transition state region (note that in this case we do not care how many times the backward piece would cross Φ_m.)

We can formally state the method outlined above by modifying Eq. (8.15) in the following way

$$k_{A \to B} = \frac{1}{Z_A} \left\langle \nabla\Phi \frac{\mathbf{p}}{m} \theta \left[\nabla\Phi \frac{\mathbf{p}}{m} \right] \delta(\Phi - \Phi_m) \chi_{A \to B}(\mathbf{p}, \mathbf{x}) \right\rangle \Bigg|_{\mathbf{p}, \mathbf{x}: \Phi[x] = \Phi_m} . \quad (8.17)$$

Here $\chi_{A \to B}(\mathbf{p}, \mathbf{x})$ is a characteristic function equal to 1 if the phase-space point $(\mathbf{p}, \mathbf{x})$ is the last crossing of $\Phi(\mathbf{x}) = \Phi_m$ by an A-to-B transition path (which, in particular, implies that it has a positive velocity $\dot{\Phi}$) and to zero otherwise.

Monte-Carlo evaluation of Eq. (8.17) usually has a tremendous advantage over the computation of the rate $k_{A \to B}$ directly from a trajectory $\Phi(t)$, because, although it does not entirely avoid simulations of the dynamics of the system, only short pieces of the trajectory, comparable in duration to transition path times, need to be generated. Starting from an intermediate point Φ_m increases the likelihood that this point belongs to a transition path and allows us to catch the system halfway through the transition, where the transition has started a short time before and will be completed soon afterwards. In contrast, had we started somewhere in A-state, it would have taken a time comparable to $\langle t_A \rangle$ before the next transition would take place, and much computational time would have been wasted on waiting for such a rare event. In fact, for many reactions of interest, $\langle t_A \rangle$ can be in a millisecond to a second range, or longer; given current computational limitations, even a single transition would then be hard or impossible to observe.

8.2.4 Transmission Coefficient and Variational Transition State Theory

The transition-state rate theory rate, Eq. (8.15), is obtained from Eq. (8.17) by replacing $\chi_{A \rightarrow B}(\mathbf{p}, \mathbf{x})$ with 1. Since the trajectories contributing to Eq. (8.17) are a subset of the ones contributing to Eq. (8.15), we immediately conclude that $k_{A \rightarrow B} \leq k_{A \rightarrow B}^{TST}$, where $k_{A \rightarrow B}^{TST}$ denotes the transition-state theory rate predicted by Eq. (8.15). Since $k_{A \rightarrow B}^{TST}$ provides an upper bound on the true rate, variational minimization of $k_{A \rightarrow B}^{TST}$, performed with respect to the mid-point Φ_m, as well as with respect to the particular function $\Phi(\mathbf{x})$ used to monitor the reaction progress, offers a useful, computationally inexpensive estimate of the rate. It immediately follows from this observation that the best TST estimate is obtained by placing Φ_m at the maximum of the potential of mean force $U(\Phi)$. This conclusion is appealing intuitively: with this choice of the mid-point Eq. (8.14) states that the transition rate is proportional to the Boltzmann probability to attain the highest (free) energy point encountered along the reaction pathway.

The difference between the exact and the TST rate coefficients is usually expressed using the *transmission coefficient* κ:

$$\kappa = k_{A \rightarrow B}/k_{A \rightarrow B}^{TST} \tag{8.18}$$

This coefficient is always less than or equal to 1. Importantly, its value is the same for the direct (A to B) and reverse (B to A) reactions, which is the reason it does not require a superscript denoting the reaction direction. This is a result of the time-reversal symmetry of the underlying dynamics, expressed by Eqs. (8.8, 8.9). Indeed, these equations imply

$$k_{A \rightarrow B}P_A^{eq} = k_{B \rightarrow A}P_B^{eq} \tag{8.19}$$

for the true rate. On the other hand, it follows from Eq. (8.14) or (8.15) that the TST rate coefficients satisfy the same condition:

$$k_{A\to B}^{TST} P_A^{eq} = k_{B\to A}^{TST} P_B^{eq} \tag{8.20}$$

Division of Eq. (8.19) by Eq. (8.20) then results in Eq. (8.18).

8.2.5 Harmonic Transition-State Theory

The transition-state theory approximation expressed by Eq. (8.15) is a no-recrossing approximation. It assumes that Φ_m is the point of no return: any molecule coming from A and crossing this point in the direction of the reaction product B proceeds to B without coming back to Φ_m. In the case where the dynamics is governed by a one-dimensional potential, as in Eq. (8.4), the optimal choice for Φ_m is at the top of the barrier separating A and B. More generally, underlying molecular dynamics is governed by a multidimensional potential energy surface $U(\mathbf{x})$, which is a function of the vector $\mathbf{x}$ of the atomic coordinates. Stable molecular conformations correspond to minima of $U(\mathbf{x})$; molecular species A and B undergo thermal fluctuations around those minima, but an occasional large fluctuation brings the molecule to the vicinity of the other minimum causing a transition between A and B. Because large energy fluctuations are suppressed by the Boltzmann distribution, a typical transition should follow a low energy path. An analogy can then be made between a molecular transition path and the path taken by a hiker crossing a mountain ridge (corresponding to the reaction barrier) situated between two mountain valleys (corresponding to A and B). In both cases a good path should lie close to a mountain pass, or a saddle point, as shown in Figure 8.5.

If only a single low-energy saddle-point separating A and B exists, we expect most transition paths to pass in the vicinity of this saddle point. To estimate the associated flux $q_{A\to B}$, we can now use the harmonic approximation in the

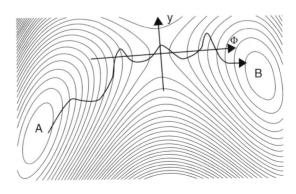

Figure 8.5 Contour plot of a two-dimensional potential energy surface, with two minima corresponding to the molecular species A and B, and a single-saddle point. Typical transition paths connecting A and B cross the barrier separating A and B in the vicinity of this saddle.

vicinity of the saddle,

$$U(\mathbf{x}) \approx U^{\ddagger} - \frac{m}{2}\omega_b^2(\Phi - \Phi^{\ddagger})^2 + \frac{m}{2}\sum_{j=1}^{n-1}(\omega_j^{\ddagger})^2 y_j^2 \qquad (8.21)$$

Here Φ, $\{y_j\}$ is a set of normal modes and $i\omega_b$, $\{\omega_j^{\ddagger}\}$ are the corresponding frequencies, of which one mode (Φ) represents the unstable direction with the imaginary frequency $i\omega_b$, such that the potential along this mode is an inverted parabola with the negative curvature $-m\omega_b^2$. The double-dagger symbol labels the saddle point and is conventional in transition state theory. As evident from the notation used in Eq. (8.21), we wish to associate the direction of the unstable mode with the reaction coordinate Φ. We will then evaluate the flux along Φ at the highest energy point along the reaction coordinate, i.e. $\Phi_m = \Phi^{\ddagger}$.

Because the potential energy described by Eq. (8.21) leads to a separable Hamiltonian, evaluation of Eq. (8.15) leads to a simple generalization of Eq. (8.14). Indeed, the multidimensional analog of Eq. (8.10) is

$$q_{A\to B} = \frac{1}{Z}\prod_{j=1}^{n-1}\int\frac{dp_j dy_j}{2\pi\hbar}e^{-\beta\frac{p_j^2}{2m}-\beta\frac{m(\omega_j^{\ddagger})^2 y_j^2}{2}}\int_0^{\infty}\frac{dp_{\Phi}}{2\pi\hbar}e^{-\beta\frac{p_{\Phi}^2}{2m}-\beta U^{\ddagger}}\frac{p_{\Phi}}{m}$$

$$= \prod_{j=1}^{n-1}z_j^{\ddagger}\frac{1}{2\pi\hbar\beta Z}e^{-\beta U^{\ddagger}} \qquad (8.22)$$

where

$$z_j^{\ddagger} = \frac{1}{\beta\hbar\omega_j^{\ddagger}} \qquad (8.23)$$

is the harmonic-oscillator partition function evaluated classically. As before, we can evaluate the rate coefficient as

$$k_{A\to B} = q_{A\to B}/P_A^{eq} = \prod_{j=1}^{n-1}z_j^{\ddagger}\frac{1}{2\pi\hbar\beta Z_A}e^{-\beta U^{\ddagger}} \qquad (8.24)$$

Finally, assuming the thermal fluctuations in the A-state around the potential minimum are not too large, we can approximate the partition function Z_A using the harmonic approximation. To this end, we expand the potential near the minimum using normal modes $\tilde{y}_j$ (note that these are generally different from y_j used in the expansion of Eq. (8.21)),

$$U(\mathbf{x}) \approx U_A + \frac{m}{2}\sum_{j=1}^{n}(\omega_j^A)^2\tilde{y}_j^2, \qquad (8.25)$$

which gives

$$Z_A \approx \prod_{j=1}^{n}\frac{1}{\beta\hbar\omega_j^A}e^{-\beta U_A} \qquad (8.26)$$

Combining Eqs. (8.23), (8.34), and (8.26), we finally obtain the central harmonic transition state theory (HTST) result:

$$k_{A \to B}^{HTST} \approx \frac{1}{2\pi} \frac{\prod\limits_{j=1}^{n} \omega_j^A}{\prod\limits_{j=1}^{n-1} \omega_j^{\ddagger}} e^{-\beta(U^{\ddagger} - U_A)} \tag{8.27}$$

In the one-dimensional case ($n=1$) this result becomes,

$$k_{A \to B}^{HTST} \approx \frac{\omega_A}{2\pi} e^{-\beta(U^{\ddagger} - U_A)}, \tag{8.28}$$

where $\omega_A = \sqrt{U_A''/m}$ is the frequency of small vibrations at the potential minimum corresponding to A and U_A'' is curvature of the potential at the minimum. The prefactor of Eq. (8.28) can be thought of as an attempt frequency, while the exponential reflects the Boltzmann probability for attaining an energy sufficient to cross the barrier.

References

1 Eyring, H. (1935). The activated complex in chemical reactions. *J. Chem. Phys.* 3: 107.

2 Pelzer, H. and Wigner, E. (1932). *Z. Phys. Chem. Abt. B* 15: 445.

3 Wigner, E. (1932). *Z. Phys. Chem. Abt. B* 19: 203.

4 Shui, V.H., Appleton, J.P., and Keck, J.C. (1972). Monte Carlo trajectory calculations of the dissociation of HCl in Ar. *J. Chem. Phys.* 56: 4266.

5 Bennett, C.H. (1977). Molecular dynamics and transition state theory: the simulation of infrequent events. In: *Algorithms for Chemical Computations*, vol. 46, 63–97. American Chemical Society.

6 Chandler, D. (1978). *J. Chem. Phys.* 68: 2959.

9

Zwanzig-Caldeiga-Leggett Model for Low-Dimensional Dynamics

In the previous Chapter, a formalism was developed to describe the transition rate in a system obeying first order kinetics, leading to a general expression for the transition rate, Eq. (8.17). In particular, we expect that the Kramers formula, Eq. (6.58), which was derived in Chapter 6 using an entirely different approach, must emerge when Eq. (8.17) is applied to the case of overdamped Langevin dynamics. We have also derived the transition-state theory approximation; for one-dimensional dynamics along a reaction coordinate, this is given by Eq. (8.28), an expression closely resembling the Kramers formula (both being examples of the Arrhenius law) but independent of the friction coefficient γ. When should one use the transition state theory, and when is the Kramers formula more appropriate? What can we say about the transition rate when neither transition state theory nor the assumption of overdamped Langevin dynamics is valid? Can we, for example, predict rate when the dynamics is described by the Generalized Langevin Equation (GLE), Eq. (1.136)? It turns out that a "unified rate theory" exists for a rather broad class of dynamical models for the coordinate x; this theory encompasses the transition-state theory and the Kramers theory, but it also extends them to account for memory effects (as in GLE) and for higher-dimensional dynamics; A powerful approach to obtaining all these theories using the same formalism is based on the observation that both the Langevin equation and the GLE can be derived from the same microscopic Hamiltonian[1-2]. This Chapter describes such a derivation and sets the stage for the next two Chapters, which will explore the role of memory and multidimensionality on reaction rates.

9.1 Low-Dimensional Models of Reaction Dynamics From a Microscopic Hamiltonian

Solving the equations motion for the entire Universe is an impossible task. We are thus forced to consider some subset of relevant degrees of freedom and treat

Molecular Kinetics in Condensed Phases: Theory, Simulation, and Analysis,
First Edition. Ron Elber, Dmitrii E. Makarov and Henri Orland.
© 2020 John Wiley & Sons Ltd. Published 2020 by John Wiley & Sons Ltd.

the rest of the world in a simplified manner. An extreme case of this approach that focuses on one wisely chosen degree of freedom x has been remarkably successful, and low-dimensional models underlie our current understanding of molecular kinetics. One- or low-dimensional models are also attractive because they offer a direct link to experimental studies: because most experimental techniques probe a single molecular property, this property (or some function of it) is a natural choice for x.

Let x be the special coordinate whose dynamics we wish to describe, and $\{x_i\}$ be the coordinates of the rest of the world (that we will refer to as *the surroundings*). Within the Newtonian mechanics, the equation of motion for the coordinate x is given by:

$$m\ddot{x} = -\partial U_{tot}(x, \{x_i\})/\partial x, \tag{9.1}$$

where U_{tot} denotes the total potential of the system and its surroundings. Since the coordinates $\{x_i\}$ are not explicitly included in the model and since the full potential $U_{tot}(x, \{x_i\})$ is unknown, we need a description of the force exerted on x by the surroundings in terms of macroscopic parameters such as, for example, the temperature. This is usually achieved through use of phenomenological models, such as the Langevin equation described in Chapter 1. Instead of treating the Langevin equation as a phenomenological description, however, it is also possible to derive it from a specially chosen microscopic potential $U_{tot}(x, \{x_i\})$[1,2]. Specifically, consider a particle with a coordinate x that is subjected to a potential $U(x)$ and coupled to a large collection (a continuum) of harmonic oscillators. The full Hamiltonian of the system (known as the Zwanzig-Caldeira-Leggett, or ZCL, Hamiltonian) is given by

$$H = \frac{p^2}{2m} + U(x) + \sum_i \frac{p_i^2}{2m} + \frac{m\omega_i^2}{2}\left(x_i - \frac{c_i x}{m\omega_i^2}\right)^2 \tag{9.2}$$

Our goal is to derive an equation for the dynamics of x alone. Before we do so, consider the equilibrium properties of x. If the combined system obeys the canonical distribution at temperature T, then the probability distribution of x is given by:

$$P(x) = C \int dx_1 dx_2 \ldots e^{-\frac{U_{tot}(x, \{x_i\})}{k_B T}} = C' e^{-\frac{U(x)}{k_B T}} \tag{9.3}$$

where C and C' are normalization constants. Therefore, the coordinate x obeys the Boltzmann distribution in a potential $U(x)$. The free energy $F(x)$ of the system is defined as

$$F(x) = -k_B T \ln P(x), \tag{9.4}$$

from which it follows that the free energy (to within an insignificant constant) is given by $U(x)$. Importantly, the potential of Eq. 9.2 contains another term,

$(x^2/2)\sum_i c_i^2/m\omega_i^2$, that depends solely on x. Had we absorbed this term in $U(x)$, the free energy would be different from $U(x)$.

To gain further insight into the physical significance of the potential $U(x)$, consider the total force acting on x,

$$f_x = -\partial U_{tot}(x, \{x_i\})/\partial x = -U'(x) + \sum_i c_i \left(x_i - \frac{c_i x}{m\omega_i^2} \right), \tag{9.5}$$

and on x_i:

$$f_j = -\partial U_{tot}(x, \{x_i\})/\partial x_j = -m\omega_j^2 \left(x_j - \frac{c_j x}{m\omega_j^2} \right) \tag{9.6}$$

If x is fixed, then the position of an oscillator x_j will fluctuate around its equilibrium value of $c_j x/m\omega_j^2$ (at which $f_j = 0$). As a result, the mean value of the second term in Eq. (9.5) is zero, and the mean value of the force along x is exactly $-U'(x)$. For this reason, the potential $U(x)$ (equal to the free energy along x) is often referred to as the *potential of mean force* (PMF).

Let us now turn to the dynamics described by the Hamiltonian of Eq. (9.2). The equations of motion of the system are:

$$m\ddot{x} = -U'(x) + \sum_i c_i \left(x_i - \frac{c_i x}{m\omega_i^2} \right) \tag{9.7}$$

$$m\ddot{x}_j = -m\omega_j^2 \left(x_j - \frac{c_j x}{m\omega_j^2} \right) \tag{9.8}$$

Given some specific trajectory $x(t)$, we can solve Eq. (9.8) to find $x_j(t)$. Indeed, this equation describes a driven harmonic oscillator in the presence of a time-dependent external force, $f_{ext}(t) = c_j x(t)$:

$$\ddot{x}_j + \omega_j^2 x_j = f_{ext}(t)/m \tag{9.9}$$

The solution of this equation is a sum of the general solution to the homogeneous equation $\ddot{x}_j + \omega_j^2 x_j = 0$ and a particular solution to Eq. 9.9. The latter can be easily found by requiring that $x_j(0) = \dot{x}_j(0) = 0$ and rewriting Eq. 9.9 in Laplace space:

$$(\lambda^2 + \omega_j^2)\hat{x}_j(\lambda) = \hat{f}_{ext}(\lambda)/m \tag{9.10}$$

Here the hat denotes the Laplace transform (e.g. $\hat{f}(\lambda) \equiv \hat{L}[f] = \int_0^\infty f(t)e^{-\lambda t}dt$). This gives:

$$\hat{x}_j(\lambda) = \hat{f}_{ext}(\lambda)/(m\lambda^2 + m\omega_j^2) = \hat{f}_{ext}(\lambda)\hat{g}(\lambda)/m$$

$$= \hat{L}\left[\int_0^t f_{ext}(t')g(t-t')dt' \right]/m, \tag{9.11}$$

in which we have recognized that a convolution of two functions becomes the product of their Laplace transforms in the Laplace space. The function $g(t)$ that we need to find has a Laplace transform given by $1/(\lambda^2 + \omega_j^2)$, and it is easy to see that $g(t) = \sin \omega_j t / \omega_j$.

Putting it all together, we write the solution in the form:

$$x_j(t) = A \cos \omega_j t + B \sin \omega_j t + c_j \int_0^t \frac{\sin \omega_j(t - t')}{m\omega_j} x_j(t')dt', \qquad (9.12)$$

where the first two terms are the general solution of the homogeneous (i.e., free oscillator) equation. In terms of the initial position, $x_j(0)$, and the initial momentum, $p_j(0) = \dot{x}_j(0)$, this can be written as

$$x_j(t) = x_j(0) \cos \omega_j t + \frac{p_j(0)}{m\omega_j} \sin \omega_j t + c_j \int_0^t \frac{\sin \omega_j(t - t')}{m\omega_j} x_j(t')dt' \qquad (9.13)$$

Integrating the last term in Eq. (9.13) by parts, we can rewrite this as

$$x_j(t) - \frac{c_j x(t)}{m\omega_j^2} = \left[x_j(0) - \frac{c_j x(0)}{m\omega_j^2} \right] \cos \omega_j t + \frac{p_j(0)}{m\omega_j} \sin \omega_j t$$

$$- \frac{c_j}{m\omega_j^2} \int_0^t \cos \omega_j(t - t')\dot{x}(t')dt' \qquad (9.14)$$

Substituting this into Eq. (9.7), we find:

$$m\ddot{x} = -U'(x) + \sum_i c_i \left[x_i(0) - \frac{c_i x(0)}{m\omega_i^2} \right] \cos \omega_i t + \sum_i c_i \frac{p_i(0)}{m\omega_i} \cos \omega_i t$$

$$- \sum_i \frac{c_i^2}{m\omega_i^2} \int_0^t \cos \omega_i(t - t')\dot{x}(t')dt' \qquad (9.15)$$

Introducing a time-dependent force

$$\xi(t) = \sum_i c_i \left[x_i(0) - \frac{c_i x(0)}{m\omega_i^2} \right] \cos \omega_i t + \sum_i c_i \frac{p_i(0)}{m\omega_i} \sin \omega_i t \qquad (9.16)$$

and a new time-dependent quantity

$$\Gamma(t) = \sum_i \frac{c_i^2}{m\omega_i^2} \cos \omega_i t, \qquad (9.17)$$

Eq. (9.15) can be further rewritten as:

$$m\ddot{x} = -U'(x) - \int_0^t \Gamma(t - t')\dot{x}(t')dt' + \xi(t). \qquad (9.18)$$

Equation (9.18) is the exact equation of motion that governs the time evolution of the selected degree of freedom x. In addition to the force $-U'(x)$, it contains two more terms. The first one depends on the velocities of the system, $\dot{x}(t)$, at different points of time. It is natural to interpret this term as viscous-type friction. The friction force, in general, depends not only on the current velocity, but also on the velocities at earlier times. In other words, Eq. (9.18) has memory.

A more conventional, memoryless case is readily recovered using a particular choice of oscillator frequencies and coupling constants. Specifically, for a continuum of oscillators we rewrite Eq. (9.17) as

$$\Gamma(t) = \int_0^\infty d\omega \rho(\omega) c^2(\omega) \cos \omega t / (m\omega^2) \tag{9.19}$$

where $\rho(\omega)d\omega$ is the number of oscillators in the frequency interval between ω and $\omega + d\omega$, and where the coupling constant c_i is assumed to be some function of frequency denoted $c(\omega)$. Using the identity $\int_0^\infty d\omega \cos \omega t = \pi \delta(t)$, we now see that if we choose the oscillators to satisfy the equation

$$c^2(\omega)\rho(\omega)/(m\omega^2) = 2\gamma \pi$$

then

$$\Gamma(t) = 2\gamma \delta(t), \tag{9.20}$$

and Eq. (9.18) becomes:

$$m\ddot{x} = -U'(x) - \gamma \dot{x} + \xi(t) \tag{9.21}$$

This equation, containing a viscous friction force $-\gamma \dot{x}$ that is proportional to the velocity, is the *Langevin equation* introduced in Chapter 1 (Eq. 1.19). Equation (9.18) is often referred to as the *generalized Langevin equation* (GLE); its overdamped limit is given by Eq. (1.136).

9.2 Statistical Properties of the Noise and the Fluctuation-dissipation Theorem

The second term in Eqs. 9.18 and 9.21 is a time-dependent force $\xi(t)$ defined by Eq. (9.16). This force depends on the initial conditions for each oscillator at $t = 0$. For sufficiently "typical" (we will return to what this means later) initial conditions, Eq. (9.16) contains many terms oscillating with different frequencies. The result is a function that will appear random in time. Since, in practice, we usually do not know the precise initial conditions of the surroundings, it makes sense, then, to adopt a statistical treatment of this force. There are two ways to proceed. In one, we imagine that Eqs. (9.16) and (9.18) are solved on a large number of computers (i.e. an "ensemble" of computers, in the language of statistical mechanics). Each computer integrates Eq. (9.18) using its own initial condition sampled randomly from some ensemble and produces its own version of the system's trajectory $x(t)$ and the fluctuating force $\xi(t)$. We then analyze how $\xi(t)$ behaves on the average, where the average is performed over the initial conditions. The second approach is to consider the average properties of $\xi(t)$ as observed from a single long trajectory, generated using one particular set of initial conditions. Will the two approaches give the same result?

9.2.1 Ensemble Approach

We start with the first approach. Suppose, for example, that all the oscillators are in thermal equilibrium, at temperature T, each obeying Maxwell-Boltzmann statistics corresponding to the initial value $x(0)$ of the coordinate x. This means that the probability distributions for each oscillator's deviation from its equilibrium position, $\Delta x_i(0) = x_i(0) - \frac{c_i x(0)}{m\omega_i^2}$, and for its initial momentum, are both Gaussian,

$$P[\Delta x_i(0)] \propto e^{-\frac{m\omega_i^2 \Delta x_i^2(0)}{2k_B T}}, \quad P[p_i(0)] \propto e^{-\frac{p_i^2(0)}{2mk_B T}}$$

with respective variances

$$\langle \Delta x_i^2(0)\rangle = \frac{k_B T}{m\omega_i^2}, \langle p_i^2(0)\rangle = mk_B T \tag{9.22}$$

It immediately follows that the force $\xi(t)$, being the sum of many terms each obeying Gaussian statistics, has itself a Gaussian distribution at any moment of time. Moreover, its mean value is zero: $\langle \xi(t)\rangle = 0$. Again, we emphasize that this is an ensemble average: if each of our computers outputs its own value of $\xi(t)$ at some moment of time, we average over the values from each computer. The variance of $\xi(t)$, $\langle \xi^2 \rangle$ is now easy to obtain from Eq. (9.16). In fact, consider a more general average of the form:

$$\langle \xi(0)\xi(t)\rangle, \tag{9.23}$$

which is the autocorrelation function of $\xi(t)$. Using Eq. (9.16), we find

$$\langle \xi(t)\xi(0)\rangle = c_i c_j \sum_{i,j} \langle \Delta x_i(0)\Delta x_j(0)\rangle \cos \omega_i t \tag{9.24}$$

If the initial displacements of each oscillator are statistically independent (i.e., $\langle \Delta x_i(0)\Delta x_j(0)\rangle = \langle \Delta x_i(0)\rangle\langle \Delta x_j(0)\rangle = 0$ for $i \neq j$) then only the terms with $i = j$ will survive in the sum of Eq. (9.24), which gives

$$\langle \xi(t)\xi(0)\rangle = k_B T \sum_i \frac{c_i^2}{m\omega_i^2} \cos \omega_i t = k_B T \Gamma(t) \tag{9.25}$$

Equation (9.25) is the *fluctuation-dissipation theorem* relating the autocorrelation function of the fluctuating force to the friction kernel $\Gamma(t)$; this result was already stated in Chapter 1 (Eq. 1.138). In the case of the memoryless friction (Eqs. (9.20–9.21)), the random force becomes delta-correlated,

$$\langle \xi(t)\xi(0)\rangle = 2k_B T \gamma \delta(t), \tag{9.26}$$

a result stated by Eq. (1.20).

9.2.2 Single-Trajectory Approach

For a specific realization of initial conditions, we could compute the force $\xi(t)$ from Eq. (9.16); If this force is monitored over a sufficiently long time interval τ, we can define the following time averages:

$$\overline{\xi} \approx \tau^{-1} \int_0^\tau \xi(t')dt'$$

$$\overline{\xi(t+t')\xi(t')} \approx \tau^{-1} \int_0^\tau \xi(t'+t)\xi(t)dt'. \tag{9.27}$$

These are counterparts of the ensemble averages $\langle \xi \rangle$, $\langle \xi(t)\xi(0) \rangle$ considered above. We hope that the resulting averages will become independent of the observation time τ when it is long enough. It is tempting to invoke the ergodicity hypothesis and to postulate that both the time and ensemble averages give the same result. Unfortunately, this is not true in general. It is easy to give a counterexample, where the time average is different from the ensemble average: just set all $\Delta x_i(0)$ and $p_i(0)$ to zero except for one oscillator. This results in a force that is strictly periodic in time, and the averages of Eq. (9.27) obviously violate the fluctuation-dissipation theorem stated by Eq. (9.25)!

We could try rescuing ergodicity by demanding that the initial condition must be in some sense typical – clearly it is unlikely to find all degrees of freedom but one in their minimum energy state. To explore this idea further, let us evaluate the time averages of Eq. (9.27) for a generic initial condition. Since the average value of a sine or a cosine is zero,

$$\overline{\sin \omega_j t} = \lim_{\tau \to \infty} \tau^{-1} \int_0^\tau \sin \omega_j t \, dt = 0 \tag{9.28}$$

we find that $\overline{\xi} = 0$ regardless of the initial conditions. To evaluate the second average from Eq. (9.27) we write

$$\xi(t'+t)\xi(t') = \sum_{i,j} \Delta x_i(0)\Delta x_j(0) \cos \omega_i(t'+t) \cos \omega_j t'$$

$$+ \sum_{i,j} m^{-2}\omega_i^{-1}\omega_j^{-1}p_i(0)p_j(0) \sin \omega_i(t'+t) \sin \omega_j t' +$$

$$+ \sum_{i,j} m^{-1}\omega_j^{-1}\Delta x_i(0)p_j(0) \cos \omega_i(t'+t) \sin \omega_j t'$$

$$+ \sum_{i,j} m^{-1}\omega_i^{-1}p_i(0)\Delta x_j(0) \sin \omega_i(t'+t) \cos \omega_j t' \tag{9.29}$$

Consider the average

$$\overline{\cos \omega_i(t'+t) \cos \omega_j t'} = \frac{1}{2}\overline{\cos(\omega_i t' + \omega_i t - \omega_j t')} + \frac{1}{2}\overline{\cos(\omega_i t' + \omega_i t + \omega_j t')} \tag{9.30}$$

Assuming that all frequencies are positive, the second term is the average of an oscillating function and is zero, while the first term is nonzero only if $\omega_i = \omega_j$, in which case this average becomes equal to $\frac{1}{2}\cos\omega_i t$. Similarly,

$$\overline{\sin\omega_i(t'+t)\sin\omega_j t'} = \frac{1}{2}\overline{\cos(\omega_i t' + \omega_i t - \omega_j t')}$$

$$-\frac{1}{2}\overline{\cos(\omega_i t' + \omega_i t + \omega_j t')} = \begin{cases} \frac{1}{2}\cos\omega_i t, \omega_i = \omega_j \\ 0, \omega_i \neq \omega_j \end{cases}, \qquad (9.31)$$

$$\overline{\sin\omega_i(t'+t)\cos\omega_j t'} = \frac{1}{2}\overline{\sin(\omega_i t' + \omega_i t + \omega_j t')}$$

$$+\frac{1}{2}\overline{\sin(\omega_i t' + \omega_i t - \omega_j t')} = \begin{cases} \frac{1}{2}\sin\omega_i t, \omega_i = \omega_j \\ 0, \omega_i \neq \omega_j \end{cases}, \qquad (9.32)$$

$$\overline{\cos\omega_i(t'+t)\sin\omega_j t'} = \frac{1}{2}\overline{\sin(\omega_i t' + \omega_i t + \omega_j t')}$$

$$-\frac{1}{2}\overline{\sin(\omega_i t' + \omega_i t - \omega_j t')} = \begin{cases} -\frac{1}{2}\sin\omega_i t, \omega_i = \omega_j \\ 0, \omega_i \neq \omega_j \end{cases} \qquad (9.33)$$

If the spectrum of oscillator frequencies is nondegenerate (i.e. $\omega_i \neq \omega_j$ for $i \neq j$), then, taking the average of Eq. (9.29) and using Eqs. (9.30–33), we obtain

$$\overline{\xi(t+t')\xi(t')} = \frac{1}{2}\sum_i c_i^2 \left[\Delta x_i^2(0) + \frac{p_i^2(0)}{m^2\omega_i^2}\right]\cos\omega_i t = \sum_i \frac{c_i^2}{m\omega_i^2}E_i(0)\cos\omega_i t, \qquad (9.34)$$

where $E_i(0)$ is the initial energy of each oscillator at $t=0$.

If the initial energy of each oscillator is exactly $k_B T$, then this result becomes identical to Eq. (9.25). There is no good justification to start with such an initial condition, however – it is just as improbable as the above example where all oscillators but one have zero energy. A more sensible approach would be to start with a random set of the oscillator positions and momenta drawn from the Boltzmann distribution. The resulting time-averaged correlation function would be a function of the initial conditions, so, at any time t, its value is a random variable. We can write is as

$$\overline{\xi(t+t')\xi(t')} \equiv C(t, \{E_j(0)\}) \qquad (9.35)$$

Since, for a Boltzmann ensemble of noninteracting harmonic oscillators, the expectation value of the initial energy is $\langle E_j(0) \rangle = k_B T$, the expectation value of $C(t, \{E_j(0)\})$ is given by

$$\langle C(t, \{E_j(0)\}) \rangle = \sum_i \frac{c_i^2}{m\omega_i^2}\langle E_i(0) \rangle \cos\omega_i t = k_B T \sum_i \frac{c_i^2}{m\omega_i^2}\cos\omega_i t \qquad (9.36)$$

which agrees with Eq. (9.25). The key argument that rescues the ergodicity hypothesis for the Langevin equation is that, in the limit when the number of oscillators N is large (i.e. the spectrum of their frequencies becomes continuous), typical deviations away from the expectation value become vanishingly small. The same argument is invoked in statistical mechanics, e.g., to justify neglecting fluctuations in the total energy of a macroscopic body that is in equilibrium with its surroundings. Consider, for example, the total initial energy $E = \sum_j E_j(0)$ of all the oscillators. Its expectation value is $Nk_B T$. Deviations from this value can be characterized by considering the quantity

$$
\begin{aligned}
\langle \Delta E^2 \rangle &= \left\langle \left(\sum_j E_j(0) - \langle E_j(0) \rangle \right)^2 \right\rangle \\
&= \sum_{ij} \langle (E_i(0) - \langle E_i(0) \rangle)(E_j(0) - \langle E_j(0) \rangle) \rangle \\
&= \sum_j \langle (E_i(0) - \langle E_i(0) \rangle)^2 \rangle
\end{aligned}
\tag{9.37}
$$

In the equation above, statistical independence of the oscillator energies necessitates that all the terms with $i \neq j$ disappear. As a result, Eq. (9.37) grows linearly with the number of oscillators, and the relative magnitude of energy fluctuations, $\langle \Delta E^2 \rangle^{1/2} / \langle E \rangle$, vanishes as the inverse square root of N.

Similarly, we can write

$$
\begin{aligned}
\langle \Delta C(t)^2 \rangle &= \langle [C(t, \{E_j(0)\}) - \langle C(t, \{E_j(0)\}) \rangle]^2 \rangle \\
&= \sum_j \langle (E_i(0) - \langle E_i(0) \rangle)^2 \rangle \frac{c_j^4}{m^2 \omega_j^4} \cos^2 \omega_j t,
\end{aligned}
\tag{9.38}
$$

and the magnitude of fluctuations of $\langle \Delta C^2(t) \rangle^{1/2}$ becomes vanishingly small, as compared to the correlation function itself, in the limit of a large number of oscillators. Therefore, the time averages obtained from a single trajectory (Eq. (9.34)) will coincide with the ensemble averages provided that the initial conditions of the oscillators are drawn from the Boltzmann distribution. As a result, the fluctuation–dissipation theorem expressed by Eqs. (9.25–9.26) applies to the *time averages* performed on the noise $\xi(t)$. For example, Eq. (9.26) now tells us that the fluctuating force $\xi(t)$ is delta-correlated in time – therefore its values at different moments of time are statistically independent and $\xi(t)$ is *white noise*. White noise can be generated numerically, and so a Langevin trajectory can be generated numerically by integrating Eq. (9.21) using an appropriate finite difference scheme, as described in Section 1.5. This, of course, is the preferred method of solving a Langevin equation – we have now reduced the dimensionality of the original problem to one degree of freedom, x, at the cost of having to use a statistical description for the remaining degrees of freedom.

The Langevin equation, Eq. (9.21), and its generalized form, Eq. (9.18), are common descriptions of the dynamics of molecules in condensed phases, especially liquids, where friction effects are important. What we have shown here is that they can be derived from a microscopic model; of course, the microscopic model employed in the derivation is not a realistic description of a molecule's surroundings in most cases. Therefore, use of the Langevin equation, say, for molecules in solution is not rigorously justified – indeed, liquid is not a collection of harmonic oscillators. Yet the Langevin equation provides a useful phenomenological model, which is guaranteed to give the correct equilibrium statistics of the degree of freedom under consideration, along with a physically meaningful description of forces exerted by the surrounding solvent. In liquids, the fluctuating force $\xi(t)$ is a result of the bombardment of the subsystem described by x by the surrounding molecules, while the friction force is often described using hydrodynamic theory. Despite being derived from a different microscopic Hamiltonian, the fluctuation–dissipation theorem is a general result; in its absence the equilibrium thermal properties of x would not be recovered.

Although the ZCL Hamiltonian, Eq. (9.2), cannot be viewed as a true microscopic description of most molecular systems, it provides a conceptual understanding of how friction/energy dissipation effects arise from conservative dynamics when only a low-dimensional subset of degrees of freedom is considered. The true power of the ZCL description will become evident in the following Chapters, where it will help us develop analytic theories for the transition rates in systems obeying GLE and Langevin dynamics.

9.3 Time-Reversibility of the Langevin Equation

The fundamental laws of physics governing temporal evolution of classical and quantum systems are time reversible. In the Newtonian mechanics, for example, if $x(t)$ is a solution of the equations of motion, then $\widetilde{x}(t) = x(-t)$ is also a solution. Since the Langevin equation can be derived from the Newtonian mechanics, we expect it to be time reversible as well. The time-reversed trajectory $\widetilde{x}(t)$, however, satisfies an odd-looking Langevin equation with a *negative* friction coefficient $-\gamma$

$$m\ddot{\widetilde{x}} = -U'(\widetilde{x}) + \gamma\dot{\widetilde{x}} + \widetilde{\xi}(t) \tag{9.39}$$

An attempt to integrate this equation using delta-correlated white noise for $\widetilde{\xi}$ will lead to a solution that will grow indefinitely. Indeed, multiplying Eq. 9.39 by $\dot{\widetilde{x}}$, we can rewrite it as

$$\frac{d}{dt}\left(\frac{m\dot{\widetilde{x}}^2}{2} + U(\widetilde{x})\right) = \gamma\dot{\widetilde{x}}^2 + \widetilde{\xi}\dot{\widetilde{x}} \tag{9.40}$$

where the first term in the right-hand side is always positive causing the energy to grow (whereas this term is negative in the original Langevin equation, causing loss of energy). The solution of this apparent paradox is that we can no longer treat $\widetilde{\xi}$ as random; rather, this force must exactly retrace the past of the fluctuating force, $\widetilde{\xi}(t) = \xi(-t)$. When it does, this time-reversed force will appear to negate the effect of the unruly negative friction in Eq. (9.40).

Physically, the necessity of time-reversing the fluctuating force is easily understood using our view of Langevin dynamics as originating from a microscopic conservative system: if we simply reverse the velocity $\dot{x}$, the resulting trajectory will not backtrack its past trajectory $x(-t)$. Rather, to obtain the exact time reversal of the dynamics it would be necessary to reverse the velocities of every degree of freedom of the surroundings, in which case the force by the surroundings on the system will also backtrack its past history.

It is instructive to explore the mathematical origin of the negative friction using the ZCL Hamiltonian, Eq. (9.2). To this end, consider the following thought experiment: Having started at $t=0$, we follow the evolution of the system and the oscillators for some time t_0, after which the velocity of the system is reversed: $\widetilde{\dot{x}}(t_0) = -\dot{x}(t_0)$ (but $\widetilde{x}(t_0) = x(t_0)$). What is the equation obeyed by the trajectory with this new initial condition, $\widetilde{x}(t)$, at $t > t_0$ if (i) the momentum of each oscillator is also reversed at $t = t_0$, i.e., $\widetilde{p}_j(t_0) = -p_j(t_0)$ and if (ii) the oscillator momenta are not reversed, $\widetilde{p}_j(t_0) = p_j(t_0)$? Note that, in this example, the reversal takes place at $t = t_0$ and not at zero time, since the time interval $0 \le t \le t_0$ was used to generate the system's past history.

To answer this question, we use Eq. (9.14) with $t = t_0$ to obtain the new initial condition after velocity reversal. We have, from Eq. (9.14),

$$\widetilde{x}_j(t_0) - \frac{c_j \widetilde{x}(t_0)}{m\omega_j^2} = \left[x_j(0) - \frac{c_j x(0)}{m\omega_j^2} \right] \cos \omega_j t_0 + \frac{p_j(0)}{m\omega_j} \sin \omega_j t_0$$
$$- \frac{c_j}{m\omega_j^2} \int_0^{t_0} \cos \omega_j (t_0 - t') \dot{x}(t') dt'$$

Differentiating this with respect to time and multiplying by mass we obtain the initial momentum

$$\widetilde{p}_j(t_0) = \pm m\omega_j \left[x_j(0) - \frac{c_j \dot{x}(0)}{m\omega_j^2} \right] \sin \omega_j t_0 \mp p_j(0) \cos \omega_j t_0$$
$$- \frac{c_j}{\omega_j} \int_0^{t_0} dt' \sin \omega_j (t_0 - t') \dot{x}(t'), \tag{9.41}$$

where the upper signs correspond to the reversed oscillator momenta, and the lower signs represent the case where the oscillators are left alone. Now we replace $x_j(0)$, $p_j(0)$ by $x_j(t_0)$, $p_j(t_0)$ in the equations of motion, Eqs. (9.13–15).

This gives, for the force exerted on x by the j-th oscillator:

$$c_j \left[\widetilde{x}_j(t) - \frac{c_j \widetilde{x}(t)}{m \omega_j^2} \right] = c_j \left[x_j(0) - \frac{c_j x(0)}{m \omega_j^2} \right] \cos \omega_j t_0 \cos \omega_j (t - t_0)$$

$$+ \frac{c_j p_j(0)}{m \omega_j} \sin \omega_j t_0 \cos(t - t_0) -$$

$$- \frac{c_j^2}{m \omega_j^2} \int_0^{t_0} \cos \omega_j (t_0 - t') \cos(t - t_0) \dot{x}(t') dt'$$

$$\pm c_j \omega_j \left[x_j(0) - \frac{c_j \dot{x}(0)}{m \omega_j^2} \right] \sin \omega_j t_0 \sin \omega_j (t - t_0) \mp$$

$$\mp c_j \frac{p_j(0)}{m \omega_j} \cos \omega_j t_0 \sin \omega_j (t - t_0) \mp \frac{c_j^2}{m \omega_j^2} \int_0^{t_0} dt' \sin \omega_j (t_0 - t') \dot{x}(t')$$

$$- \frac{c_j^2}{m \omega_j^2} \int_0^{t_0} dt' \sin \omega_j (t_0 - t') \sin \omega_j (t - t_0) \dot{x}(t') -$$

$$- \frac{c_j^2}{m \omega_j^2} \int_{t_0}^{t} \cos \omega_j (t_0 - t') \widetilde{x}(t') dt' \tag{9.42}$$

Using trigonometric identities, this can be transformed into

$$c_j \left[\widetilde{x}_j(t) - \frac{c_j \widetilde{x}(t)}{m \omega_j^2} \right] = c_j \left[x_j(0) - \frac{c_j x(0)}{m \omega_j^2} \right] \cos[\omega_j t_0 \mp \omega_j (t - t_0)]$$

$$+ \frac{c_j p_j(0)}{m \omega_j} \sin[\omega_j t_0 \mp \omega_j (t - t_0)] -$$

$$- \frac{c_j^2}{m \omega_j^2} \int_0^{t_0} \cos[\omega_j (t_0 - t') \mp \omega_j (t - t_0)] \dot{x}(t') dt'$$

$$- \frac{c_j^2}{m \omega_j^2} \int_{t_0}^{t} \cos \omega_j (t - t') \widetilde{x}(t') dt' \tag{9.43}$$

The total force exerted by the surroundings on x is Eq. 9.43 summed over j. Consider first the case (i) where the momenta of all degrees of freedom are reversed at $t = t_0$. The first two terms in the r.h.s. of Eq. 9.43, summed over all the oscillators will then give $\widetilde{\xi}(t) = \xi[t_0 - (t - t_0)]$, which is the time-reversed fluctuating force. The last two terms in the r.h.s. of Eq. (9.43) summed over all j's will yield a friction-like force. Let us write it explicitly for the memoryless case (where $\sum_j c_j^2 \cos \omega_j t / m \omega_j^2 = 2 \gamma \delta(t)$):

$$-2\gamma \int_0^{t_0} \delta(2t_0 - t - t')\dot{x}(t')dt' - 2\gamma \int_0^{t_0} \delta(t - t')\dot{\tilde{x}}(t')dt'$$

$$= -2\gamma\dot{x}[t_0 - (t - t_0)] - \gamma\dot{\tilde{x}}(t)$$

Note that the second term is an integral over half of the delta function, which is why the factor of two has disappeared. But since $\tilde{x}(t) = x[t_0 - (t - t_0)]$, is the time-reversed trajectory, the above two terms have opposite signs and add up to a friction term with a flipped sign, $+\gamma\dot{\tilde{x}}(t)$, resulting in a time-reversed Langevin equation with a negative friction coefficient, $m\ddot{\tilde{x}} = -U'(\tilde{x}) + \gamma\dot{\tilde{x}} + \xi[t_0 - (t - t_0)]$.

Consider now the second case, where the momentum associated with x is reversed, but the oscillator momenta are not. In this case Eq. (9.43) becomes

$$c_j\left[\tilde{x}_j(t) - \frac{c_j\tilde{x}(t)}{m\omega_j^2}\right] = c_j\left[x_j(0) - \frac{c_jx(0)}{m\omega_j^2}\right]\cos[\omega_j t] + \frac{c_jp_j(0)}{m\omega_j}\sin[\omega_j t] -$$

$$- \frac{c_j^2}{m\omega_j^2}\int_0^{t_0}\cos[\omega_j(t - t')]\dot{x}(t')dt' - \frac{c_j^2}{m\omega_j^2}\int_{t_0}^t\cos\omega_j(t - t')\dot{\tilde{x}}(t')dt'$$

$$(9.44)$$

The fluctuating force $\xi(t)$ continues to evolve according to Eq. (9.16), while the friction force resulting from the last two terms of Eq. (9.44) becomes, in the memoryless case, simply $-\gamma\dot{\tilde{x}}(t)$ (note that, in fact, only the last term contributes) – the friction coefficient is positive!

References

1 Caldeira, A.O. and Leggett, A.J. (1983). Quantum tunneling in a dissipative system. *Ann. Phys.* 149: 374.

2 Zwanzig, R. (2001). *Nonequilibrium Statistical Mechanics*. Oxford University Press.

10

Escape from a Potential Well in the Case of Dynamics Obeying the Generalized Langevin Equation: General Solution Based on the Zwanzig-Caldeira-Leggett Hamiltonian

10.1 Derivation of the Escape Rate

In 1941, Kramers proposed that a chemical reaction in a polyatomic molecule can be described using the model where one degree of freedom is treated explicitly while the effect of the remaining degrees of freedom is described using the theory of Brownian motion [1]. Specifically, those extra degrees of freedom exert a force that includes a deterministic, velocity-dependent friction as well as a random component. Kramers then found the transition rate for the model where the system escapes out of a one-dimensional potential well, as illustrated in Fig. 10.1. The solution to the problem given by Kramers and described in Chapter 6 relied on the assumption that the friction force is proportional to the velocity; as shown in the previous Chapter, however, the friction force may, in general, depend not only on the instantaneous velocity, but also on the velocity values at earlier moments of time. Here we examine the more general case of the dynamics described by the Generalized Langevin Equation (GLE)

$$m\ddot{x} = -U'(x) - \int_0^t \Gamma(t - t')\dot{x}(t')dt' + \xi(t), \tag{10.1}$$

where the memory kernel is related to the Gaussian-distributed noise, $\xi(t)$, through the fluctuation–dissipation relationship, $\langle \xi(t)\xi(0) \rangle = k_B T\Gamma(t)$. Our goal is to estimate the rate coefficient for the escape from a metastable well (state A) described by a potential $U(x)$, whose shape is depicted in Figure 10.1. The final state (B) is not depicted there because its presence does not affect the result of our calculation.

Two methods of solving this problem were described in Chapter VI for the case where Eq. (10.1) reduces to the ordinary Langevin equation in the over-damped limit, but neither method can be easily extended to the GLE case. The method presented below follows the ingenuous and elegant approach due to Pollak [2] and is based on the equivalence of the GLE dynamics and conservative dynamics of an extended system where the coordinate x is coupled to a

Molecular Kinetics in Condensed Phases: Theory, Simulation, and Analysis,
First Edition. Ron Elber, Dmitrii E. Makarov and Henri Orland.
© 2020 John Wiley & Sons Ltd. Published 2020 by John Wiley & Sons Ltd.

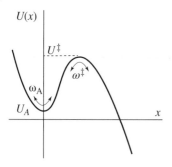

Figure 10.1 The problem of escape from a one-dimensional potential well.

bath of harmonic oscillators with an appropriate choice of oscillator frequencies and coupling coefficients (Chapter 9). The total multidimensional potential describing the system (cf. Eq. (9.2)),

$$U(x) + \sum_i \frac{m\omega_i^2}{2}\left(x_i - \frac{c_i x}{m\omega_i^2}\right)^2,$$

(10.2)

has a minimum corresponding to the initial state (A) and a saddle corresponding to the reaction barrier. In the Langevin/Kramers view, the transition-state theory's no-recrossing assumption (Chapter 8) is violated because of thermal noise that can kick the system in either direction along x – a system crossing the free energy barrier in the x direction will not necessarily be committed to the product state B. From the point of view of the multidimensional potential of Eq. (10.2), recrossings arise because of the coupling of x to other degrees of freedom, making x a poor choice for the reaction coordinate. If the harmonic approximation is used in the vicinity of the multidimensional saddle on this potential, it is possible to introduce a reaction coordinate Φ that is "perfect" (inasmuch as the harmonic approximation holds) because its dynamics is decoupled from that of other degrees of freedom. When the system crosses the barrier in the direction of this coordinate, it will proceed straight to the product B without recrossings. This coordinate is simply the normal mode corresponding to the unstable direction at the saddle. The escape rate is then simply estimated as the harmonic transition state theory (HTST) rate for the potential of Eq. (10.2), as described in Section 8.2.5.

To employ HTST, we diagonalize the matrix of the second derivatives (also known as the Hessian matrix) of our potential:

$$\mathbf{h} = \begin{pmatrix} U'' + \sum_j \frac{c_j^2}{m\omega_j^2} & -c_1 & -c_2 & \dots \\ -c_1 & m\omega_1^2 & 0 & 0 \\ -c_2 & 0 & m\omega_2^2 & 0 \\ \dots & 0 & 0 & \dots \end{pmatrix}$$

(10.3)

If we denote its eigenvalue by $m\lambda^2$, and if $\mathbf{I}$ is the identity matrix, we have

$$\det(\mathbf{h} - \lambda^2\mathbf{I}) = \left(U'' + \sum_j \frac{c_j^2}{m\omega_j^2} - m\lambda^2 \right) \prod_j (m\omega_j^2 - m\lambda^2)$$
$$- \sum_j c_j^2 \prod_{i \neq j} m(\omega_i^2 - \lambda^2) = 0,$$

or, equivalently,

$$\left(U'' + \sum_j \frac{c_j^2}{m\omega_j^2} - m\lambda^2 \right) = \frac{1}{m} \sum_j \frac{c_j^2}{\omega_j^2 - \lambda^2} \tag{10.4}$$

The reaction coordinate we seek is associated with an imaginary eigenvalue, $\lambda = i\mu$, and the second derivative of the potential should be evaluated at the top of the barrier, where $U'' = -m(\omega^{\ddagger})^2$, $\omega^{\ddagger}$ being the upside down barrier frequency. Eq. (10.4) can then be rewritten as

$$[-m(\omega^{\ddagger})^2 + m\mu^2] = -\frac{1}{m} \sum_j \frac{c_j^2 \mu^2}{\omega_j^2(\omega_j^2 + \mu^2)} \tag{10.5}$$

Recalling that the memory kernel in Eq. (10.1) is given by $\Gamma(t) = \sum_j \frac{c_j^2}{m\omega_j^2} \cos \omega_j t$ (see Eq. (9.17)), Eq. (10.5) can be rewritten in the form

$$\mu^2 - (\omega^{\ddagger})^2 + \mu\hat{\Gamma}(\mu)/m = 0, \tag{10.6}$$

where $\hat{\Gamma}(\mu)$ denotes the Laplace transform of the memory kernel.

We are now ready to apply the HTST expression, Eq. (8.27), to our problem. We rewrite it in the following form:

$$k_{A \to B} \approx \frac{1}{2\pi m^{1/2}} \left(\frac{\det \mathbf{H}_A}{\det' \mathbf{H}^{\ddagger}} \right)^{1/2} e^{-\beta(U^{\ddagger} - U_A)}, \tag{10.7}$$

where $\mathbf{H}_A$ and $\mathbf{H}^{\ddagger}$ is the Hessian matrix (Eq. (10.3)) evaluated, respectively, at the potential minimum and at the saddle, and where the prime indicates that the negative eigenvalue $-m\mu^2$ corresponding to the unstable mode is omitted from the determinant. Using Eq. (10.3) we find

$$\det \mathbf{H} = \left(U'' + \sum_j \frac{c_j^2}{2m\omega_j^2} \right) \prod_j m\omega_j^2 - \sum_j c_j^2 \prod_{i \neq j} m\omega_i^2 = U'' \prod_j m\omega_j^2 \tag{10.8}$$

Since we have $U'' = m\omega_A^2$ for the potential curvature at the reactant state minimum (Fig. 10.1), we can write

$$\frac{\det \mathbf{H}_A}{\det' \mathbf{H}^{\ddagger}} = \frac{\det \mathbf{H}_A}{\det \mathbf{H}^{\ddagger}}(-m\mu^2) = \frac{m\omega_A^2}{m(\omega^{\ddagger})^2} m\mu^2.$$

Substituting this into Eq. (10.7), we obtain

$$k_{A \to B} \approx \frac{\omega_A}{2\pi} \frac{\mu}{\omega^{\ddagger}} e^{-\beta(U^{\ddagger} - U_A)} \tag{10.9}$$

If, instead of applying HTST to the multidimensional system described by the potential of Eq. (10.2), we use it for the one-dimensional system described by the potential $U(x)$, the result $k_{A \to B}^{TST}$ will be given by Eq. 8.28. We, therefore, write

$$k_{A \to B} = k_{A \to B}^{TST} \kappa, \tag{10.10}$$

with the transmission coefficient given by the ratio of two frequencies,

$$\kappa = \frac{\mu}{\omega^{\ddagger}} \tag{10.11}$$

The result expressed by Eqs. (10.10) and (10.11) was originally derived (in a different way) by Grote and Hynes [3, 4] and is known as the Grote-Hynes formula. The quantity μ can be interpreted as the "reactive frequency", i.e. the frequency of the barrier along the "true" reaction coordinate that coincides with the unstable normal coordinate in the high-dimensional space of Eq. (10.2).

10.2 The Limit of Kramers Theory

It is instructive to consider the simplest example of a GLE with an exponentially decaying memory kernel,

$$\Gamma(t) = \Gamma(0)e^{-t/\tau_c}, \tag{10.12}$$

where τ_c is a characteristic memory time. To obtain this kernel from the expression for the memory kernel, Eq. (9.17), we need to introduce a continuum of harmonic oscillators. Let $\rho(\omega)d\omega$ be the number of oscillators with frequencies between ω and $\omega + d\omega$, and let the coupling coefficient c_j be some function of its frequency, $c_j = c(\omega_j)$. Then Eq. (9.17) becomes an integral,

$$\Gamma(t) = \int_0^{\infty} d\omega \frac{c^2(\omega)\rho(\omega)}{m\omega^2} \cos \omega t \tag{10.13}$$

One can verify directly that choosing

$$\frac{c^2(\omega)\rho(\omega)}{m\omega^2} = \frac{2}{\pi \tau_c} \frac{1}{\omega^2 + \tau_c^{-1}} \Gamma(0) \tag{10.14}$$

yields the desired friction kernel, Eq. (10.12). To determine the reactive frequency μ, we now rewrite Eq. (10.5) as

$$[-m(\omega^{\ddagger})^2 + m\mu^2] = -\frac{1}{m} \int_0^{\infty} d\omega \frac{c^2(\omega)\rho(\omega)\mu^2}{\omega^2(\omega^2 + \mu^2)} = -\Gamma(0) \frac{|\mu|}{|\mu| + \tau_c^{-1}} \tag{10.15}$$

Since both the l.h.s. and the r.h.s. of this equation are even functions of μ, we can assume $\mu > 0$ without loss of generality. Before discussing the general case, let us examine the behavior of the solution of Eq. (10.15) in the limit where it satisfies the condition $\mu \ll \tau_c^{-1}$. Neglecting the term cubic in μ, we write

$$\mu^2 + \frac{\xi(0)\tau_c}{m}\mu - (\omega^{\ddagger})^2 \approx 0 \tag{10.16}$$

The positive root of this equation is

$$\mu = \sqrt{(\omega^{\ddagger})^2 + \frac{\gamma^2}{4m^2}} - \frac{\gamma}{2m}, \tag{10.17}$$

where we set

$$\gamma = \hat{\Gamma}(0) = \Gamma(0)\tau_c \tag{10.18}$$

In combination with Eq. (10.9), this constitutes the result derived, in a different way, by Kramers [1] and described in Chapter 6.

The same solution for the reactive frequency μ can also be obtained directly from Eq. (10.6) by assuming Langevin dynamics and thus taking $\Gamma(t) = 2\gamma\delta(t)$. This shows that the Kramers formula is recovered when the memory time τ_c is so short that the friction kernel can be approximated by the delta function,

$$\Gamma(t) \approx 2\delta(t) \int_0^{\infty} \Gamma(t)dt \tag{10.19}$$

Of special interest is the overdamped limit, where $\gamma/m \gg \omega^{\ddagger}$ and Eq. (10.17) can be approximated by

$$\mu \approx m(\omega^{\ddagger})^2/\gamma \tag{10.20}$$

In this limit, the transmission coefficient (Eq. (10.11)) becomes

$$\kappa \approx \frac{m\omega^{\ddagger}}{\gamma} \ll 1 \tag{10.21}$$

Combined with Eq. (10.10), this gives the overdamped limit of the Kramers formula, which was derived in Chapter 6 (Eq. (6.58)). This regime is usually deemed relevant for chemical reactions in solution, and, in particular, for biochemical processes. If the main source of friction comes from the hydrodynamic forces exerted by the solvent, the friction coefficient γ is proportional to the solvent viscosity, and so the reaction rate is inversely proportional to the viscosity, in accord with our intuition that reactions in more viscous solvents should proceed slower.

In the opposite limit, $\gamma/m \ll \omega^{\ddagger}$, we have $\kappa \to 1$, and the transition-state theory result $k_{A \to B}^{TST}$ is recovered. Transition-state theory is, however, known to break down again at small values of friction – this case, known as the "energy diffusion regime", will not be considered here; its discussion can be found in several excellent reviews [5, 6].

10.3 Significance of Memory Effects

The Kramers result amounts to neglecting memory effects and assuming Markovian dynamics along the coordinate x. Inclusion of the memory effects, however, leads to a host of new phenomena, which we will now proceed to examine using the model of exponential memory, Eq. (10.12).

Let us rewrite Eq. (10.15) for the reactive frequency in the following way:

$$\mu^2 + \widetilde{\gamma} \frac{\mu}{\mu \tau_c + 1} - (\omega^{\ddagger})^2 = 0, \tag{10.22}$$

where the rescaled friction coefficient

$$\widetilde{\gamma} = \gamma/m = \Gamma(0)\tau_c/m \tag{10.23}$$

has the units of inverse time, and where only the positive roots of this equation are to be considered. In the limit $\widetilde{\gamma} \to 0$ the transition state theory result is still recovered, as in the Markov case.

The limit $\widetilde{\gamma} \to \infty$ is more interesting. If the friction coefficient γ (or $\widetilde{\gamma}$) and the memory time τ_c are regarded as independent parameters and if the latter is fixed while the former is increased then, again, the overdamped limit of the Kramers theory, Eq. (10.20), is attained. In many relevant applications, however, γ and τ_c are interdependent. If, for example, the solution viscosity is increased, this may slow down both the reaction of interest and the solvent's dynamics itself. If both γ and τ_c are proportional to the solvent viscosity, then the ratio

$$\nu^2 = \frac{\widetilde{\gamma}}{\tau_c} \tag{10.24}$$

remains constant. This ratio has the units of frequency squared, and will be taken now as an independent parameter. We then ask: if this parameter is fixed, how will the solution to Eq. (10.22) behave in the limit $\gamma \to \infty$, $\tau_c \to \infty$?

Rewriting Eq. (10.22) one more time in the following form,

$$\mu^3 \tau_c + \mu^2 + \mu \tau_c [\nu^2 - (\omega^{\ddagger})^2] - (\omega^{\ddagger})^2 = 0 \tag{10.25}$$

and keeping in mind that the solution μ that we seek must be positive, we find that two distinct scenarios are encountered depending on the sign of the difference $\nu^2 - (\omega^{\ddagger})^2$.

If $\nu^2 > (\omega^{\ddagger})^2$, we find

$$\mu \approx \frac{(\omega^{\ddagger})^2}{\nu^2 - (\omega^{\ddagger})^2} \frac{1}{\tau_c} = \frac{(\omega^{\ddagger})^2}{\widetilde{\gamma}} \frac{1}{1 - (\omega^{\ddagger})^2/\nu^2} \to 0 \tag{10.26}$$

Comparing this with Eq. (10.20), we observe that, as in the no-memory case, the transition rate displays a pseudo-Kramers behavior where it is still

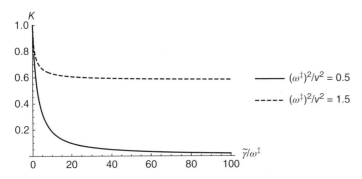

Figure 10.2 Grote-Hynes theory with exponentially decaying friction kernel: The transmission coefficient is plotted as a function of the friction coefficient for the two regimes defined by the value of the parameter v. When v is below the barrier frequency, the transmission coefficient decays to a finite value, but when v exceeds the barrier frequency, the transmission coefficient decays to zero as the friction coefficient is increased.

inversely proportional to the friction coefficient, but with a different (larger) proportionality factor.

If $v^2 < (\omega^\ddagger)^2$, however, Eq. (10.26) predicts a reactive frequency that has the wrong sign. The correct, positive solution remains nonzero as $\gamma \to \infty$, attaining a finite value

$$\lim_{\gamma \to \infty} \mu = \sqrt{(\omega^\ddagger)^2 - v^2},\tag{10.27}$$

and resulting in a transition-state-theory-like result with a friction-independent rate,

$$\lim_{\gamma \to \infty} k_{A \to B} = \frac{\omega_A}{2\pi} \sqrt{1 - v^2/(\omega^\ddagger)^2}\, e^{-\beta(U^\ddagger - U_A)}\tag{10.28}$$

The behavior of the transmission coefficient $\kappa = \mu/\omega^\ddagger$ obtained by solving the cubic equation Eq. (10.25) is shown in Figure 10.2 as a function of the friction coefficient.

What is the physical significance of the frequency parameter v? Why two distinct physical regimes depending on whether it is greater or smaller than the barrier frequency $\omega^\ddagger$? We will return to these questions in the next Chapter.

10.4 Applications of the Kramers Theory to Chemical Kinetics in Condensed Phases, Particularly in Biomolecular Systems

Many condensed phase chemical reactions are characterized by energy landscapes that are too complex to permit a microscopic description. Even when such a description is possible, it often does not provide immediate insight into the behavior of the experimental observable (i.e. the quantity Φ introduced in Chapter 8), which may be a complex, generally nonlinear function of the atomic coordinates. Although the dynamics of Φ may be quite complex, a phenomenological description in which this observable itself is assumed to obey Langevin dynamics (Eq. (8.4)) has been used extensively to rationalize experimental observations and/or predict rates of complex molecular transformations. In the language of the Kramers problem, the coordinate x is identified with the experimental observable Φ, $x \equiv \Phi$.

At the very least, equilibrium distribution of Φ predicted by the dynamical model should match the equilibrium distribution $P_{eq}(\Phi)$ observed experimentally or in simulations. Since the Langevin equation is consistent with the Bolzmann statistics, this necessitates that

$$P_{eq}(\Phi) \propto e^{-\beta U(\Phi)} \tag{10.29}$$

or

$$U(\Phi) = \beta^{-1} \ln P_{eq}(\Phi) + \text{constant}, \tag{10.30}$$

which means that $U(\Phi)$ must be the *free energy* as a function of Φ. We also call this function the potential of mean force (PMF), as its derivative gives the average force acting along the coordinate Φ when its value is fixed (see Section 9.1). A further simplification that is often employed is the neglect of the inertial term $(m\ddot{\Phi})$ in the Langevin equation, which means that the dynamics is overdamped (cf. Section 1.3). If so, then using Eqs. (10.10) and (10.21), we can write the rate of a barrier-crossing transition in the form

$$k_{A \to B} \approx \frac{\sqrt{U_A'' |U_{TS}''|}}{2\pi\gamma} e^{-\beta(U^{\ddagger} - U_A)}, \tag{10.31}$$

where $U_A''(\Phi) = m\omega_A^2$ is the PMF curvature at the reactant minimum and $U_{TS}'' = -m(\omega^{\ddagger})^2$ the curvature at the barrier top. Note that Eq. (10.31) does not contain the mass m. If the equilibrium distribution of Φ is known, then the friction coefficient is the only dynamical parameter entering into the expression for the transition rate. Thus experimental data can be interpreted using a single adjustable parameter γ or, equivalently, a diffusion coefficient $D = k_B T/\gamma$.

Curiously, of the many seminal contributions that Kramers has made in physics, his formula describing the rate of escape may now be the most impactful one, since the biophysics and biochemistry community has embraced (perhaps too optimistically) this result as a universal description of biochemical kinetics. Because most biochemical transitions take place in aqueous environments, the Langevin equation provides a useful phenomenological description of solvent friction, which tends to slow biochemical transitions down relative to their transition-state theory rates.

10.5 A Comment on the Use of the Term "Free Energy" in Application to Chemical Kinetics and Equilibrium

When the precise definition of the terms "free energy of the transition state" and "free energy of the reactant/product" is not specified, their usage can lead to confusion and even to physically incorrect conclusions. For example, consider the quantity chemists refer to as the equilibrium constant. For the unimolecular reaction defined by Eq. (8.1), this is the ratio of the equilibrium populations of B and A,

$$K = \frac{P_{eq}^B}{P_{eq}^A} = \frac{Z_B}{Z_A} = \frac{k_{A \to B}}{k_{B \to A}} \tag{10.32}$$

Using the usual definition of the free energy of a state A (or B),

$$F_{A(B)} = -k_B T \ln Z_{A(B)}, \tag{10.33}$$

one relates the equilibrium constant to the free energy difference $\Delta F = F_B - F_A$ between the states A and B,

$$K = \exp\left(-\frac{\Delta F}{k_B T}\right), \tag{10.34}$$

an expression familiar to everyone who has taken a freshman chemistry class. But confusion may arise when one uses, e.g. Equation 10.31 (and the analogous expression for the rate of the reverse process), which gives

$$K \approx \sqrt{\frac{U_A''}{U_B''}} \exp\left(-\frac{U_B - U_A}{k_B T}\right). \tag{10.35}$$

The problem is that the potential of mean force $U(\Phi)$ is also referred to as a free energy (see above), and so $U_B - U_A$ is the free energy difference between the B and A minima. The difference between these two formulas for the same equilibrium constant arises from the difference in how each "free energy" is defined. To clarify this, let us start with the general definition of the free energy of a state s, $F(s)$, which can be related to the equilibrium probability $P_{eq}(s)$ of being in this state, which, in turn, is the sum of the equilibrium probabilities of all phase space points (or quantum states in the quantum case) consistent with s:

$$e^{-\beta F(s)} = P_{eq}(s) = \sum_{\mathbf{x}, \mathbf{p} \in s} P(\mathbf{x}, \mathbf{p}).$$

The potential of mean force (Eq. (10.30)), as a function of the reaction coordinate Φ, is the free energy associated with a precise value of Φ. In contrast, the free energies F_A and F_B appearing in Eqs. (10.33–34) are associated with conformations in which Φ fluctuates around its most probable values (i.e. the minima of $U(\Phi)$). A minimum with a smaller U'' results in greater fluctuations and, therefore, a larger value of the partition function Z (and thus a lower free energy). When using Eq. (10.34) (or Eq. (10.35)), it is thus important to remember how the free energies appearing in these equations are defined.

References

1 Kramers, H.A. (1940). Brownian motion in a field of force and the diffusion model of chemical reactions. *Physica* 7: 284–304.

2 Pollak, E. (1986). Theory of activated rate-processes - a new derivation of Kramers expression. *J. Chem. Phys.* 85 (2): 865–867.

3 Grote, R.F. and Hynes, J.T. (1980). The stable states picture of chemical reactions. II. Rate constants for condensed and gas phase reaction models. *J. Chem. Phys.* 73 (6): 2715–2732.

4 Grote, R.F. and Hynes, J.T. (1981). Reactive modes in condensed phase reactions. *J. Chem. Phys.* 74 (8): 4465–4475.

5 Hanggi, P., Talkner, P., and Borkovec, M. (1990). 50 years after Kramers. *Rev. Mod. Phys.* 62: 251.

6 Pollak, E. and Talkner, P. (2005). Reaction rate theory: what it was, where is it today, and where is it going? *Chaos* 15 (2): 26116.

11

Diffusive Dynamics on a Multidimensional Energy Landscape

11.1 Generalized Langevin Equation with Exponential Memory can be Derived from a 2D Markov Model

Not all phenomena in molecular kinetics can be adequately described within a one-dimensional framework. In this chapter, we will explore chemical kinetics governed by a Langevin equation with more than one degree of freedom. It will be seen that such a description can sometimes be further reduced to one degree of freedom and dealt with using the methods developed in the previous chapter. Even then, the multidimensional description has its advantages, both computational and conceptual. In particular, it will provide physical insight into some of the results of the Grote-Hynes theory described in Chapter 10.3.

We start with a simple model containing two degrees of freedom, x and y, with the (free) energy landscape given by:

$$U_{tot}(x, y) = U(x) + \frac{1}{2}K(y - x)^2, \tag{11.1}$$

The system is governed by the Langevin equations of motion

$$m\ddot{x} = -\gamma_x \dot{x} - \partial U_{tot}/\partial x + \zeta_x(t) = -\gamma_x \dot{x} - U'(x) - K(x - y) + \xi_x(t)$$

$$0 = -\gamma_y \dot{y} - \partial U_{tot}/\partial y + \zeta_y(t) = -\gamma_y \dot{y} - K(y - x) + \xi_y(t) \tag{11.2}$$

Here the random noise components $\xi_{x,y}$ obey Gaussian distributions with zero means, are statistically independent (i.e., $\langle \xi_x(t)\xi_y(t') \rangle = 0$ for any t, t'), and satisfy the fluctuation dissipation theorem:

$$\langle \xi_{x(y)}(t)\xi_{x(y)}(t') \rangle = 2k_B T \gamma_{x(y)} \delta(t - t'). \tag{11.3}$$

Overdamped limit is assumed for the dynamics along y, and so the inertial term proportional to $\ddot{y}$ is omitted; we will eventually assume overdamped dynamics for x as well, omitting the $m\ddot{x}$ term in the first equation. Before doing so, however, we derive the equation of motion for x alone in the more general case

Molecular Kinetics in Condensed Phases: Theory, Simulation, and Analysis,
First Edition. Ron Elber, Dmitrii E. Makarov and Henri Orland.
© 2020 John Wiley & Sons Ltd. Published 2020 by John Wiley & Sons Ltd.

with inertial effects present. To do so, write the second of Equations (11.2) in the form

$$\dot{y} + \frac{1}{\tau_y} y = \frac{1}{\tau_y} x(t) + \frac{\xi_y(t)}{\gamma_y} \equiv f(t) \tag{11.4}$$

This equation describes the motion of an overdamped harmonic oscillator, with a relaxation time

$$\tau_y = \gamma_y / K, \tag{11.5}$$

driven by a force $f(t)$. The solution is

$$\begin{aligned}
y(t) &= y(0)e^{-t/\tau_y} + \int_0^t e^{-(t-t')/\tau_y} f(t') dt' \\
&= y(0)e^{-t/\tau_y} + \frac{1}{\tau_y} \int_0^t e^{-(t-t')/\tau_y} x(t') dt' + \frac{1}{\gamma_y} \int_0^t e^{-(t-t')/\tau_y} \xi_y(t') dt' \\
&= y(0)e^{-t/\tau_y} + x(t) - x(0)e^{-t/\tau_y} - \int_0^t e^{-(t-t')/\tau_y} \dot{x}(t') dt' \\
&\quad + \frac{1}{\gamma_y} \int_0^t e^{-(t-t')/\tau_y} \xi_y(t') dt',
\end{aligned} \tag{11.6}$$

where integration by parts was used to obtain the last line. Substituting this back into the first line of Eq. (11.2), we find

$$\begin{aligned}
m\ddot{x} = -\gamma_x \dot{x} - K \int_0^t e^{-(t-t')/\tau_y} \dot{x}(t') dt' - U'(x) \\
+ K[y(0) - x(0)]e^{-t/\tau_y} + \xi_x(t) + \frac{K}{\gamma_y} \int_0^t e^{-(t-t')/\tau_y} \xi_y(t') dt' \tag{11.7}
\end{aligned}$$

Eq. (11.7) can be interpreted as a generalized Langevin equation of the form:

$$m\ddot{x} = -U'(x) - \int_0^t \Gamma(t - t') \dot{x}(t') dt' + \xi(t) \tag{11.8}$$

with a memory kernel

$$\xi(t) = 2\gamma_x \delta(t) + K \exp\left(-\frac{t}{\tau_y}\right) \tag{11.9}$$

and with an external random force

$$\begin{aligned}
\xi(t) &= \xi_x(t) + \tilde{\xi}_y(t) + \xi_i(t) \\
\tilde{\xi}_y(t) &= \frac{K}{\gamma_y} \int_0^t e^{-(t-t')/\tau_y} \xi_y(t') dt' \\
\xi_i(t) &= K[y(0) - x(0)]e^{-t/\tau_y}. \tag{11.10}
\end{aligned}$$

This force satisfies the fluctuation–dissipation theorem, Eq. (9.25). To show this, let us first note that the three force terms in Eq. (11.10) are statistically independent. This allows us to write the force autocorrelation function in the form:

$$\langle \xi(t)\xi(0) \rangle = \langle \tilde{\xi}_x(t)\tilde{\xi}_x(0) \rangle + \langle \tilde{\xi}_y(t)\tilde{\xi}_y(0) \rangle + K^2 \langle [y(0) - x(0)]^2 \rangle e^{-\frac{t}{\tau_y}}$$

From the definition of $\tilde{\xi}_y(t)$ in Eq. (11.10), we have $\tilde{\xi}_y(0) = 0$, and thus the second term in the above sum vanishes. To evaluate the remaining expression, we perform averages both over time and over the initial distribution of the coordinate $y(0)$, which is assumed to be the Boltzmann distribution $P[y(0)] \propto \exp\{-\beta U_{tot}[x(0), y(0)]\}$, with the given initial condition $x(0)$ (cf. Section 9.2.1, where the same assumption was used to derive the GLE from a microscopic Hamiltonian). Because the potential of Eq. (11.1) is quadratic with respect to y, this is a Gaussian distribution, and application of the equipartition theorem immediately leads to $K\langle [y(0) - x(0)]^2 \rangle / 2 = k_B T / 2$. We finally arrive at the following expression for the autocorrelation function of the force:

$$\langle \xi(t)\xi(0) \rangle = 2k_B T \gamma_x \delta(t) + K k_B T e^{-\frac{t}{\tau_y}} = k_B T \Gamma(t) \tag{11.11}$$

Indeed, Eq. (11.11) has the form of the fluctuation-dissipation theorem (cf. (Eq. 9.25)). If the friction coefficient γ_x is negligible then x obeys a generalized Langevin equation with exponential memory kernel considered in Section 10.2. Therefore, non-Markov dynamics governed by a GLE with an exponential memory kernel can be obtained as a projection of two-dimensional Markov dynamics on just one degree of freedom. In other words, Markov behavior is restored by considering one extra degree of freedom explicitly [1–3]. This result leads to a useful simulation trick: while solving the stochastic integro-differential GLE (Eq. (11.8)) numerically is tricky, solving a two-dimensional Langevin equation with no memory is straightforward. Moreover, this approach can be generalized to the case of nonexponential memory kernels: If we couple the coordinate of interest, x, to more than one overdamped harmonic oscillator, it will obey a GLE whose memory kernel is a sum of exponentials; to the extent that the memory kernel of interest can be approximated as a sum of decaying exponentials, then, we have a general numerical recipe for solving GLEs.

Let us now return to the problem of escape from a potential which is well described by a potential $U(x)$ depicted in Figure 10.1. In Section 10.2, we have analyzed this problem under the assumption that the dynamics of x can be described using a GLE with exponential memory, Eq. (10.12). We now see that this GLE can be obtained if the coordinate x is coupled to another degree of freedom y, if the intrinsic friction coefficient γ_x can be neglected, and if the following relationships hold between the parameters of the model discussed here and the GLE parameters introduced in Sections 10.2–3 (see Eqs. (10.18) and (10.24)):

$$\tau_c = \tau_y, \quad \gamma = \gamma_y, \quad \nu^2 = K/m \tag{11.12}$$

The problem from Section 10.2 is thus mapped onto that of escape from a *two-dimensional* potential well described by Eq. (11.1), where the dynamics in the extended xy-space has no memory (Eq. (11.2)). This equivalent formulation of the problem offers an intuitive explanation to some of the findings from Section 10.3. Specifically, consider the shape of the potential $U_{tot}(x, y)$ in the vicinity of the maximum of $U(x)$ (taken to be at $x = 0$); to second order in both coordinates we have

$$U_{tot}(x, y) \approx -m(\omega^{\ddagger})^2 x^2/2 + mv^2(x - y)^2/2, \tag{11.13}$$

where the barrier frequency is defined as $(\omega^{\ddagger})^2 = -U_0''(0)/m$ (see Figure 11.1). Recall that, in Section 10.3, two distinct regimes were encountered. When $v > \omega^{\ddagger}$, the rate was found to behave as $1/\gamma$ at large values of the friction coefficient, while for $v < \omega^{\ddagger}$ the rate approached a friction-independent limit as $\gamma \to \infty$. In terms of the present two-dimensional model, these findings are easy to understand. We can think of Eq. (11.13) as describing two particles connected by a spring of stiffness $K = mv^2$ (Figure 11.1). The x-particle is subjected to a potential $U(x) \approx -m(\omega^{\ddagger})^2 x^2/2$ but no friction, while the y-particle is subjected to friction. When the x-particle is trying to cross the barrier, it has to drag the y-particle along.

In the limit $\gamma = \gamma_y \to \infty$ the dynamics along y becomes very slow. If $v < \omega^{\ddagger}$, then the x-particle can cross the saddle ($x{=}0$, $y{=}0$) without waiting for the y-particle to move. The effective potential seen by the coordinate x is then $U(x, 0) = -m[(\omega^{\ddagger})^2 - v^2]x^2$, which is a parabolic barrier with a renormalized barrier frequency $\sqrt{(\omega^{\ddagger})^2 - v^2}$. Since there is no friction along x, the transition rate becomes friction-independent in this case.

When $v > \omega^{\ddagger}$, however, the potential $U_{tot}(x, 0)$ has a minimum rather than a maximum at $x{=}0$. This means that a transition cannot happen unless both x and y are changed together. The two particles must move in a concerted fashion, and thus the friction acting on the y-particle slows the transition down. In fact, when the spring connecting the two particles becomes infinitely stiff ($v^2 \to \infty$), the two particles must move together as a united system ($x = y$), which, consequently, experiences a friction force equal to $-\gamma_y \dot{y} = -\gamma \dot{x}$. The problem is then reduced to the high friction limit of the one-dimensional Kramers problem, precisely as found in Section 10.3.

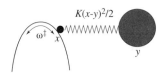

$K(x{-}y)^2/2$

$\omega^{\ddagger}$ x

y

Figure 11.1 Equation (11.1) describes two particles tethered together with a linear spring. In the vicinity of the barrier, the energy landscape of this system is quadratic.

11.2   Theory of Multidimensional Barrier Crossing

Let us know focus on the overdamped case with respect to both x and y, where Eq. (11.2) becomes

$$\gamma_x \dot{x} = -U'(x) - K(x - y) + \xi_x(t)$$
$$\gamma_y \dot{y} = -K(y - x) + \xi_y(t) \tag{11.14}$$

In the vicinity of the barrier (assumed to be located at $x = y = 0$), we write

$$U'(x) \approx U''(x)x \equiv -K_0 x, \tag{11.15}$$

where $K_0 = -U''(0) = m(\omega^{\ddagger})^2$ is the barrier curvature.
If averaged over thermal noise, Eq. 11.14 becomes

$$\gamma_x \langle \dot{x} \rangle = K_0 \langle x \rangle - K(\langle x \rangle - \langle y \rangle)$$
$$\gamma_y \langle \dot{y} \rangle = -K(\langle y \rangle - \langle x \rangle), \tag{11.16}$$

with the noise term disappearing and with the angular brackets describing the average over the noise. This is a system of linear differential equations, which can be written in a more general form [4]

$$\mathbf{\Gamma} \frac{d}{dt} \begin{pmatrix} \langle x \rangle \\ \langle y \rangle \end{pmatrix} = -\mathbf{H}^{\ddagger} \begin{pmatrix} \langle x \rangle \\ \langle y \rangle \end{pmatrix}, \tag{11.17}$$

where

$$\mathbf{\Gamma} = \begin{pmatrix} \gamma_x & 0 \\ 0 & \gamma_y \end{pmatrix} \tag{11.18}$$

is a friction matrix, and

$$\mathbf{H}^{\ddagger} = \begin{pmatrix} \partial^2 U_{tot}/\partial x^2 & \partial^2 U_{tot}/\partial x \partial y \\ \partial^2 U_{tot}/\partial y \partial x & \partial^2 U_{tot}/\partial y^2 \end{pmatrix} \Bigg|_{x=y=0} = \begin{pmatrix} -K_0 + K & -K \\ -K & K \end{pmatrix} \tag{11.19}$$

is the Hessian matrix (the matrix of the 2nd derivatives of U_{tot}). Rewriting Eq. (11.17) as

$$\frac{d}{dt} \begin{pmatrix} \langle x \rangle \\ \langle y \rangle \end{pmatrix} = -\mathbf{\Gamma}^{-1} \mathbf{H}^{\ddagger} \begin{pmatrix} \langle x \rangle \\ \langle y \rangle \end{pmatrix} \tag{11.20}$$

we see that its solution this can be written in the form

$$\begin{pmatrix} \langle x \rangle \\ \langle y \rangle \end{pmatrix} = \sum_{i=1,2} a_i \mathbf{e}_i \exp(\mu_i t), \tag{11.21}$$

where μ_i and $\mathbf{e}_i$ are, respectively, the eigenvalues and the eigenvectors of $-\boldsymbol{\Gamma}^{-1}\mathbf{H}^{\ddagger}$, and where a_i are some coefficients. Using Eqs. (11.18) and (11.19), one finds the quadratic characteristic equation for μ's:

$$\mu_{1,2}^2 + \mu_{1,2}\left(\frac{K}{\gamma_y} + \frac{K - K_0}{\gamma_x}\right) - \frac{KK_0}{\gamma_x\gamma_y} = 0 \tag{11.22}$$

Of particular interest is its positive root, which leads to an exponentially growing solution:

$$\mu_1 = -\frac{1}{2}\left(\frac{K}{\gamma_y} + \frac{K - K_0}{\gamma_x}\right) + \frac{1}{2}\sqrt{\left(\frac{K}{\gamma_y} + \frac{K - K_0}{\gamma_x}\right)^2 + 4\frac{KK_0}{\gamma_x\gamma_y}} \tag{11.23}$$

Now let us pause and recall that an alternative way of looking at the dynamics governed by Eq. (11.2) is to consider the motion of the coordinate x alone, which is governed by a GLE with the friction kernel described by Eq. (11.9). In this case we can use the Grote-Hynes equation, Eq. (10.6), to determine the reactive frequency μ. Using $(\omega^{\ddagger})^2 = K_0/m$ we obtain from Eq. (10.6)

$$\mu^2 - \frac{K_0}{m} + \mu\hat{\Gamma}(\mu) = 0 \tag{11.24}$$

In the overdamped limit, the term μ^2 (originating from the acceleration term in the Langevin equation) should be neglected, which gives $\mu\hat{\Gamma}(\mu) = K_0$. Finally, taking the Laplace transform of Eq. (11.9),

$$\hat{\Gamma}(\mu) = \gamma_x + \frac{K}{\mu + K/\gamma_y},$$

we find that the Grote-Hynes reactive frequency obeys Eq. (11.22), whose positive solution is given by Eq. (11.23)!

In the two-dimensional view of the Grote-Hynes problem with exponential memory that we have just developed, we have two decoupled modes, whose directions in the x-y space are given by the vectors $\mathbf{e}_1$ and $\mathbf{e}_2$. The first of the two modes is unstable, with a displacement along this mode growing exponentially, as $\exp(\mu_1 t)$. The dynamics along this mode is analogous to one-dimensional overdamped dynamics in a parabolic barrier. Indeed, consider overdamped Langevin dynamics in a one-dimensional parabolic barrier described by a potential $U(x) = -m(\omega^{\ddagger})^2 x^2/2$, with a friction coefficient γ_x; the coordinate x (averaged over thermal noise) evolves according to the one-dimensional Langevin equation

$$\gamma_x\langle\dot{x}\rangle = m(\omega^{\ddagger})^2\langle x\rangle,$$

which gives and exponentially growing solution of the form $\langle x\rangle \propto e^{\mu t}$ with $\mu = m(\omega^{\ddagger})^2/\gamma_x$. The growth rate μ is, in this case, is identical to the reactive frequency determined from the high friction limit of the Kramers theory (cf. Eq. (10.20)).

Since the dynamics along $\mathbf{e}_1$ is memoryless, and since it is decoupled from the dynamics along $\mathbf{e}_2$, we should be able to compute the transition rate in the two-dimensional potential of Eq. (11.1) from the Kramers theory using this mode as the new reaction coordinate. Although by doing this, for this specific case, we will merely recover the result already obtained from the Grote-Hynes theory (as will be seen below), a generalization of this method to different potentials and to problems of higher dimensionality leads to a powerful new method. Specifically, let $\mathbf{x}$ be the vector describing the system's configuration and let $U(\mathbf{x})$ be the system's potential energy. The dynamics of the system is described by

$$\Gamma\frac{d\mathbf{x}}{dt} = -\nabla U + \boldsymbol{\xi}(t), \tag{11.25}$$

where $\boldsymbol{\xi}(t)$ is a vector of random forces satisfying the appropriate fluctuation dissipation relationships and Γ is a friction matrix. We now assume that the potential U has an "A" minimum corresponding to the reactant, a single transition-state saddle, and a product "B" minimum. In the vicinity of the reactant minimum the potential can be approximated by its quadratic expansion,

$$U \approx U_A + \frac{1}{2}(\mathbf{x} - \mathbf{x}_A)^T \mathbf{H}_A(\mathbf{x} - \mathbf{x}_A), \tag{11.26}$$

and, similarly, in the vicinity of the saddle (taken to be located at the origin, $\mathbf{x}^{\ddagger} = 0$) we have

$$U \approx U^{\ddagger} + \frac{1}{2}\mathbf{x}^T \mathbf{H}^{\ddagger}\mathbf{x} \tag{11.27}$$

We now wish to write the rate coefficient $k_{A \to B}$ as a product of its (harmonic) transition-state theory value (Eq. (8.27) in Section 8.2.5) and a transmission coefficient κ. To find the latter, we employ the Kramers theory, with the reaction coordinate aligned with the special unstable mode $\mathbf{e}_1$; as above, this mode is an eigenvector of $-\Gamma^{-1}\mathbf{H}^{\ddagger}$ with a positive eigenvalue equal to $\mu \equiv \mu_1$.

More precisely, we choose a reaction coordinate Φ to coincide (locally, in the vicinity of the saddle) with the projection of $\mathbf{x}$ onto the direction of the vector $\mathbf{e} \equiv \mathbf{e}_1$ (Figure 11.2); the subscript "1" will be omitted from now on to simplify the notation. According to this choice, all points belonging to a hyperplane that is perpendicular to $\mathbf{e}$ have the same value of Φ. Therefore, the value of the reaction coordinate, as a function of x, can be written as

$$\Phi(\mathbf{x}) = \mathbf{e}\mathbf{x}, \tag{11.28}$$

with the convention that Φ becomes zero at the transition-state saddle. Note that the reaction coordinate does not have to be given by Eq. (11.28) globally, since, for example, Eq. (11.28) may poorly describe the physics of the problem in the vicinity of the reactant-state minimum (Figure 11.2).

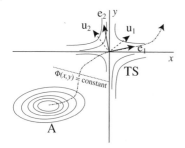

Figure 11.2 Mapping of the problem defined by two-dimensional Langevin dynamics onto a one-dimensional Kramers problem. The dash–dotted line specifies a reaction coordinate $\Phi(x, y)$ by requiring that all points lying on this line (or a hyperplane in a higher-dimensional space) have the same value of Φ. The reaction coordinate is locally aligned with the eigenvector of the matrix $-\mathbf{\Gamma}^{-1}\mathbf{H}^{\ddagger}$ corresponding to the growth mode with a positive eigenvalue μ_1, such that $\Phi \approx \mathbf{e}_1\mathbf{x}$ in the vicinity of the transition-state (TS) saddle. The eigenvectors of $-\mathbf{\Gamma}^{-1}\mathbf{H}^{\ddagger}$ are generally different from the normal modes $\mathbf{u}_{1,2}$, which are the eigenvectors of $\mathbf{H}^{\ddagger}$.

To apply the Kramers formula, we need to determine the shape of the potential of mean force near the barrier top. To do so, we introduce a set of normal coordinates $x_n = \mathbf{x}\mathbf{u}_n$, where $\mathbf{u}_n$ are the eigenvectors of $\mathbf{H}^{\ddagger}$. In terms of the normal coordinates, Eq. (11.27) can be rewritten as

$$U \approx U^{\ddagger} + \frac{m}{2} \sum_n (\omega_n^{\ddagger})^2 x_n^2, \tag{11.29}$$

where one of the oscillation frequencies $\omega_n^{\ddagger}$ is imaginary, since the expansion is written near a saddle. Equation (11.28) can also be written in terms of the components e_n of the vector $\mathbf{e}$ in this new coordinate system,

$$\Phi(\mathbf{x}) = \sum_n e_n x_n \tag{11.30}$$

The potential of mean force $U_{\Phi}(\Phi)$ corresponding to the reaction coordinate Φ is determined by the expression

$$e^{-\beta U_{\Phi}(\Phi)} = e^{-\beta U^{\ddagger}} \int \prod_n \frac{dp_n dx_n}{2\pi\hbar} e^{-\beta\left[\frac{p_n^2}{2m} + \frac{m(\omega_n^{\ddagger})^2 x_n^2}{2}\right]} \delta\left(\sum_n x_n e_n - \Phi\right), \tag{11.31}$$

where p_n are the momenta associated with the normal coordinates. The delta function in the above integral enforces that integration is constrained to the hyperplane with the same value of the reaction coordinate. Without this delta-function, Eq. (11.31) would yield a product of harmonic-oscillator partition functions (the integral along the mode with imaginary frequency would, however, diverge). The delta-function constraint couples the integration variables, so that Eq. (11.31) is no longer a simple product of one-dimensional integrals.

This difficulty can be circumvented using a trick: writing the delta-function using the identity

$$\delta\left(\sum_n x_n e_n - \Phi\right) = \frac{1}{2\pi}\int_{-\infty}^{\infty} ds\, e^{is\left(\sum_n x_n e_n - \Phi\right)}, \tag{11.32}$$

allows us to express Eq. (11.31) as

$$e^{-\beta U_\Phi(\Phi)} = e^{-\beta U^\ddagger}\int_{-\infty}^{\infty} ds\, e^{-is\,\Phi}\prod_n I_n(s). \tag{11.33}$$

Here

$$\begin{aligned}
I_n(z) &= \iint \frac{dp_n dx_n}{2\pi\hbar}\, e^{-\beta\left[\frac{p_n^2}{2m}+\frac{m(\omega_n^\ddagger)^2 x_n^2}{2}\right]+is\,x_n e_n}\\
&= \iint \frac{dp_n dx_n}{2\pi\hbar}\, e^{-\beta\left[\frac{p_n^2}{2m}+\frac{m(\omega_n^\ddagger)^2}{2}\left(x_n-\frac{is\,e_n}{\beta m(\omega_n^\ddagger)^2}\right)^2\right]-\frac{e_n^2 s^2}{2\beta m(\omega_n^\ddagger)^2}}\\
&= z_n^\ddagger e^{-\frac{e_n^2 s^2}{2\beta m(\omega_n^\ddagger)^2}}, \tag{11.34}
\end{aligned}$$

where $z_n^\ddagger = \frac{k_B T}{\hbar\omega_n^\ddagger}$ is the classical partition function of a harmonic oscillator. Finally, performing the Gaussian integral in Eq. (11.33) we arrive at

$$\begin{aligned}
e^{-\beta U_\Phi(\Phi)} &= e^{-\beta U^\ddagger}\prod_n z_n^\ddagger \int_{-\infty}^{\infty} ds\, e^{-is\Phi-\frac{s^2}{2\beta m}\sum_n \frac{e_n^2}{(\omega_n^\ddagger)^2}}\\
&= e^{-\beta U^\ddagger}\prod_n z_n^\ddagger \sqrt{\frac{m\beta}{2\pi(\omega_\Phi^\ddagger)^2}}\, e^{-\beta\frac{m}{2}(\omega_\Phi^\ddagger)^2\Phi^2}, \tag{11.35}
\end{aligned}$$

where we have introduced the barrier frequency for the potential of mean force measured along the reaction coordinate Φ.

$$(\omega_\Phi^\ddagger)^2 = \left(\sum_n e_n^2/(\omega_n^\ddagger)^2\right)^{-1} \tag{11.36}$$

The reader has probably noticed that when one of the frequencies $\omega_n^\ddagger$ is imaginary, as is the case for a saddle, then the integral in Eq. (11.35) diverges, and, moreover, the associated partition function z_n is a complex number! If we nevertheless disregard these observations boldly, the final result for the shape of the barrier measured along the coordinate Φ turns out to be correct:

$$U_\Phi(\Phi) = \text{constant} + \frac{m(\omega_\Phi^\ddagger)^2}{2}\Phi^2 = \text{constant} - \frac{m|\omega_\Phi^\ddagger|^2}{2}\Phi^2. \tag{11.37}$$

One can verify this by taking ω_1 to be imaginary and by assuming that the reaction coordinate is aligned with the associated normal coordinate (i.e., $e_1 = 1$, $e_n = 0$ for $n \neq 1$). In fact, we *expect* that $\omega_\Phi^\ddagger$ is imaginary, because the direction **e** is downhill from the top of the barrier ensuring that that the potential of Eq. (11.37) has a barrier and allowing us to interpret $|\omega_\Phi^\ddagger|$ as the barrier frequency.

It is sensible to choose the reaction coordinate Φ as some smooth line that passes through the reactant (A) minimum and through the transition-state saddle, and that is aligned with the mode **e** in the vicinity of the barrier; we then can write an equation analogous to Eq. (11.35) to describe the potential of mean force in the vicinity of the minimum,

$$e^{-\beta U_\Phi(\Phi \approx \Phi_A)} \approx e^{-\beta U_A} \prod_n z_n^A \sqrt{\frac{m\beta}{2\pi(\omega_\Phi^A)^2}} e^{-\beta \frac{m}{2}(\omega_\Phi^A)^2(\Phi - \Phi_A)^2}, \tag{11.38}$$

where Φ_A is the value of the reaction coordinate at the minimum, $m(\omega_\Phi^A)^2$ is the potential of mean force's curvature at the minimum, and z_n^A are the partition functions for each normal mode at the minimum. Dividing Eq. (11.38) by Eq. (11.35), we find that the free energy barrier is given by

$$e^{-\beta[U_\Phi(0) - U_\Phi(\Phi_A)]} = e^{-\beta(U^\ddagger - U_A)} \frac{\left|\prod_n z_n^\ddagger\right|}{\prod_n z_n^A} \frac{|\omega_\Phi^\ddagger|}{\omega_\Phi^A} = e^{-\beta(U^\ddagger - U_A)} \frac{\prod_n \omega_n^A}{\left|\prod_n \omega_n^\ddagger\right|} \frac{|\omega_\Phi^\ddagger|}{\omega_\Phi^A}$$

$$= e^{-\beta(U^\ddagger - U_A)} \left(\frac{\det \mathbf{H}_A}{|\det \mathbf{H}^\ddagger|}\right)^{1/2} \frac{|\omega_\Phi^\ddagger|}{\omega_\Phi^A} \tag{11.39}$$

Finally, applying the Kramers formula to one-dimensional dynamics along Φ and recalling that the transmission factor is given by $\kappa = \mu/|\omega_\Phi^\ddagger|$, (where, as before, μ is the positive eigenvalue of $-\boldsymbol{\Gamma}^{-1}\mathbf{H}^\ddagger$), we obtain

$$k_{A \to B} = \frac{\omega_\Phi^A}{2\pi} \kappa e^{-\beta[U_\Phi(0) - U_\Phi(\Phi_A)]} = \frac{\mu}{2\pi} \left(\frac{\det \mathbf{H}_A}{|\det \mathbf{H}^\ddagger|}\right)^{1/2} e^{-\beta(U^\ddagger - U_A)} \tag{11.40}$$

The precise choice of the reaction coordinate near the reactant state minimum (which affects the value of ω_Φ^A) does not affect the outcome of this calculation. Moreover, the mass m does not enter into Eq. (11.40), as should be expected in the overdamped case.

Eq. (11.40) is the central prediction of the theory of multidimensional barrier crossing developed by Langer [5], and it is known as the Langer formula. The derivation described here is, however, different from Langer's and is closely related to the theory proposed by Berezhkovskii and Szabo [4].

The schematics of Figure 11.2 and the idea of using the unstable mode $\mathbf{e}_1$ as a reaction coordinate near the saddle is supported by numerical simulations, as shown in Figure 11.3. Indeed, this Figure illustrates the alignment of the typical transition path with the direction of this mode.

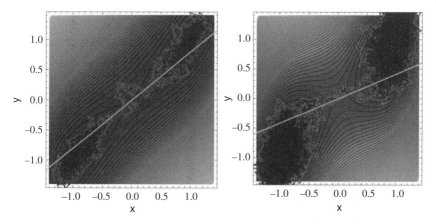

Figure 11.3 Numerically simulated trajectories undergoing Brownian dynamics in a potential of Eq. (11.1) (with $U(x)$ being a symmetric double well, which is the same in both cases). The spring constant K is lower in the case depicted on the right and the friction coefficients are the same for x and y, $\gamma_x = \gamma_y$. The green line shows the direction of the unstable mode $\mathbf{e}_1$. (*See color plate section for color representation of this figure*).

11.3 Breakdown of the Langer Theory in the Case of Anisotropic Diffusion: the Berezhkovskii-Zitserman Case

There is an interesting (and practically important – see below) case where the Langer theory does not work. This is the case of so-called anisotropic diffusion, where the friction coefficient for some degree of freedom (e.g. y) is much greater than for others. This case is illustrated in Figure 11.4, which shows dynamics in the same potential as that in right panel of Figure 11.3, with the only difference being that the friction coefficient along y was taken to be 50 times that along x, $\gamma_y = 50\gamma_x$.

How many transitions do we see in Figure 11.4? Answering this question requires careful thought, as it may depend on what we call a transition. The dynamics along y is now slow, and the system may undergo multiple large fluctuations along x before it settles within a basin of attraction corresponding to the reactant or the product. Such fluctuations are, however, not noticeable in the dynamics of the slow mode y.

Another important feature of the trajectory shown in Figure 11.4 is that transition paths may now easily bypass the saddle. Because our derivation of the Langer theory assumes that the reaction coordinate goes through the saddle, this observation immediately suggests that the Langer theory may not work in this case. Indeed, this turns out to be true; in particular, for the two-dimensional potential of Eq. (11.1) this breakdown occurs when the spring stiffness K is sufficiently low.

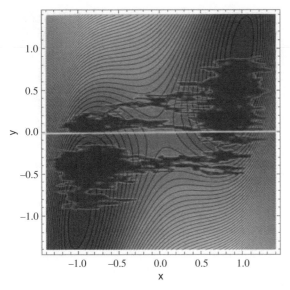

Figure 11.4 Numerically simulated trajectories undergoing Brownian dynamics for the same potential as in the right panel of Figure 11.3, but with the friction coefficient along y increased 50-fold, i.e., $\gamma_y = 50\gamma_x$. The green line shows the direction of the unstable mode $\mathbf{e}_1$. (*See color plate section for color representation of this figure*).

To demonstrate the breakdown of the Langer theory, let us first examine its predictions in the limit $\gamma_y/\gamma_x \to \infty$. Using γ_x/γ_y as a small parameter and expanding Eq. (11.23) in a Taylor series, we obtain:

$$
\begin{aligned}
\mu_1 &\approx -\frac{K-K_0}{2\gamma_x} + \frac{1}{2}\sqrt{\left(\frac{K-K_0}{\gamma_x}\right)^2 + 2\frac{(K-K_0)k}{\gamma_y\gamma_x} + 4\frac{KK_0}{\gamma_x\gamma_y}} \\
&= -\frac{K-K_0}{2\gamma_x} + \frac{|K-K_0|}{2\gamma_x}\sqrt{1 + \frac{2K\gamma_x}{\gamma_y}\frac{(K+K_0)}{(K-K_0)^2}} \\
&\approx -\frac{K-K_0}{2\gamma_x} + \frac{|K-K_0|}{2\gamma_x} + \frac{K\gamma_x}{2\gamma_y}\frac{(K+K_0)}{(K-K_0)} + \dots
\end{aligned}
\tag{11.41}
$$

If $K > K_0$, the first two terms of Eq. (11.41) cancel each other out, and, to lowest order, μ_1 is inversely proportional to γ_y. As a result, the Langer rate, Eq. (11.40), is also inversely proportional to γ_y. In terms of the scheme depicted in Figure 11.1, when the spring connecting the two coupled particles is stiff, both of them move together during a transition, and the overall friction coefficient is close to the largest of the two friction coefficients, γ_y.

When the connecting spring is soft (i.e. $K < K_0$), however, the lowest order term in Eq. 11.41 is independent of γ_y and is equal to

$$
\mu_1 \approx \frac{K_0 - K}{\gamma_x}
\tag{11.42}
$$

This means that the rate predicted by the Langer theory is independent of γ_y. More precisely, using Eq. (11.19) for the Hessian (and a similar equation written for the reactant state A), Eq. (11.40) gives, in this case,

$$k_{A \to B} = \frac{1}{2\pi} \frac{K_0 - K}{\gamma_x} \left(\frac{K_A}{K_0} \right)^{1/2} e^{-\beta(U^\ddagger - U_A)}, \tag{11.43}$$

where $K_A = (\partial^2 U / \partial x^2)_{x=x_A}$ is the molecular stiffness in the reactant state. In the limit $K \ll K_0$ this further reduces to the overdamped limit of the Kramers formula for the one-dimensional potential $U(x)$.

The independence of the result of the friction coefficient γ_y can be understood as follows: When the motion along y becomes very slow, the value of y stays virtually unchanged in the course of each individual transition path. In the vicinity of the transition-state saddle, then, the potential experienced by the system as x changes is simply $-K_0 x^2 / 2 + K x^2 / 2 = (K - K_0) x^2 / 2$. This change in the effective barrier curvature from $-K_0$ to $-(K_0 - K)$ explains the renormalization of the reactive frequency μ from its Kramers value $\mu = K_0 / \gamma_x$ expected in the absence of coupling between x and y to the result of Eq. (11.42).

Berezhkovskii and Zitserman [6, 7] realized that the independence of the reaction rate of the friction coefficient γ_y predicted by the Langer theory for $K < K_0$ cannot hold indefinitely as the value of γ_y is increased. In the case of strongly anisotropic friction, where $\gamma_y \gg \gamma_x$, the reaction rate must eventually become inversely proportional to γ_y. Indeed, when it is much slower than the dynamics along x, the dynamics along y must become the rate limiting bottleneck for the transition rate. In fact, it is easy to estimate the true limit of the transition rate $k_{A \to B}$ achieved in the case $\gamma_y / \gamma_x \gg 1$. When fluctuations along x become much faster than the motion along y, the instantaneous force experienced by the slow y-degree of freedom can be replaced by the value obtained by averaging this force over the fluctuations in x. As discussed before, this average force can be expressed as the derivative of the potential of mean force $U_{eff}(y)$ defined as

$$e^{-\beta U_{eff}(y)} = \int_{-\infty}^{\infty} dx \, e^{-\beta U(x,y)} \tag{11.44}$$

The motion along y is thus effectively described by a Langevin equation in a potential $U_{eff}(y)$ and with a friction coefficient γ_y. Consequently, $k_{A \to B}$ can simply be approximated by the Kramers theory prediction for this effective potential. If the minimum (corresponding to A) and the maximum of this potential are located at y_{min} and y_{max}, then the result is

$$k_{A \to B}(\gamma_y \to \infty) = \frac{[U_{eff}''(y_{min})]^{1/2} |U_{eff}''(y_{max})|^{1/2}}{2\pi\gamma_y} e^{-\beta[U_{eff}(y_{max}) - U_{eff}(y_{min})]} \tag{11.45}$$

which is inversely proportional to the friction coefficient along y. The more general case of moderately large friction is quite complicated and, in fact, no analytical solution to this problem had been known until recently. This case will not be discussed here in any detail, but let us point out the following simple, semi-empirical relation [8] that was found to interpolate between the Langer case and the limit of Eq. 11.45 and to work well for any γ_y:

$$k_{A \to B} \approx \left(\frac{1}{k_{A \to B}^{(L)}} + \frac{1}{k_{A \to B}(\gamma_y \to \infty)} \right)^{-1}. \tag{11.46}$$

Here $k_{A \to B}^{(L)}$ is the Langer theory estimate given by Eq. (11.40).

The case of anisotropic friction considered by Berezhkovskii and Zitserman has attracted recent attention in the context of mechanical pulling experiments performed on individual molecules [9, 10]. In those experiments, a mechanical force is exerted on a molecule via a setup where this molecule is attached to a much larger force probe. A crude sketch of such a setup can be provided by Figure 11.1 if we think of x as the mechanical extension of the molecule and of y as the position of the force probe. The spring connecting the two particles further provides a primitive model for a linker (typically a polymer chain) connecting the molecule and the probe. A common example of a force probe is a dielectric bead placed in a laser optical trap: by manipulating the trap the experimentalist can exert a mechanical force on the molecule, and by watching the bead move the experimentalist tries to deduce the molecular motion (i.e. the time evolution of x). When the molecule undergoes a conformational rearrangement, the value of x (and, consequently, the value of the experimental observable y) changes. In the case of first-order, two-state kinetics we expect the time evolution of y to resemble the trajectory shown in Figure 8.2 of Chapter 8, and so the transition rates between the two states can be measured. But does the transition rate measured in this way coincide with the intrinsic rate in the absence of the probe, or does the coupling of the molecule to the probe change the observed rate?

If we use our two-dimensional model as a description of the dynamics of the two coupled degrees of freedom, that of the molecule (x) and the probe (y), then much insight can be obtained into the above question using the results we already have. If the Langer theory holds and if the linker is soft $(K \ll K_0)$ then Eq. (11.43) tells us that the rate observed in the presence of the probe would be close to that without the probe (i.e. the intrinsic rate for the potential $U(x)$). The Berezhkovskii-Zitserman theory, however, is bound to diminish one's optimism: If the probe (and thus the corresponding friction coefficient γ_y) is large enough, the observed rate would be controlled by the hydrodynamic drag on the probe, as predicted by Eq. (11.45).

References

1 Straub, J.E. and Berne, B.J. (1986). Energy diffusion in many-dimensional Markovian systems: the consequences of competition between inter- and intramolecular vibrational energy transfer. *J. Chem. Phys.* 85: 2999–3006.

2 Straub, J.E., Borkovec, M., and Berne, B.J. (1986). Non-Markovian activated rate processes: comparison of current theories with numerical simulation data. *J. Chem. Phys.* 84: 1788–1794.

3 Zwanzig, R. (2001). *Nonequilibrium Statistical Mechanics.* Oxford University Press.

4 Berezhkovskii, A. and Szabo, A. (2005). One-dimensional reaction coordinates for diffusive activated rate processes in many dimensions. *J. Chem. Phys.* 122 (1): 14503.

5 Langer, J.S. (1969). *Ann. Phys.* (N.Y.) 54: 258.

6 Berezhkovskii, A. and Zitserman, V.Y. (1989). *Chem. Phys. Lett.* 158: 369.

7 Berezhkovskii, A.M. and Zitserman, V.Y. (1991). Activated Rate Processes in the multidimensional case. Consideration of recrossings in the multidimensional Kramers problem with anisotropic friction. *Chem. Phys.* 157: 141–155.

8 Berezhkovskii, A.M., Szabo, A., Greives, N. et al. (2014). Multidimensional reaction rate theory with anisotropic diffusion. *J. Chem. Phys.* 141 (20): 204106.

9 Nam, G.M. and Makarov, D.E. (2016). Extracting intrinsic dynamic parameters of biomolecular folding from single-molecule force spectroscopy experiments. *Protein Sci.* 25: 123–134.

10 Cossio, P., Hummer, G., and Szabo, A. (2015). On artifacts in single-molecule force spectroscopy. *Proc. Natl. Acad. Sci. U. S. A.* 112 (46): 14248–53.

12

Quantum Effects in Chemical Kinetics

12.1 When is a Quantum Mechanical Description Necessary?

Most of the discussion so far has rested on the assumption that the motion of molecules and atoms could be described by Newton's second law. Molecules and atoms are, however, microscopic objects, which obey laws of quantum mechanics. In practice, laws of classical mechanics can be used instead of quantum mechanics if the quantum de Broglie length, which is a measure of delocalization of a quantum particle, is much shorter than the characteristic length scale of the problem. For an atom of mass m moving with a velocity v, the de Broglie wavelength is given by

$$\lambda = \frac{2\pi\hbar}{mv} \tag{12.1}$$

where $\hbar \approx 1.05 \times 10^{-34} J \times s$ is Planck's constant. Let v be a typical thermal velocity of the atom. This can be estimated by demanding that the atom's kinetic energy, $mv^2/2$, be comparable to the thermal energy $k_B T$, which gives (to within a numerical factor) $v \sim \sqrt{k_B T/m}$. For a hydrogen atom at $T = 300K$, then, Eq. (12.1) gives $\lambda \sim 2.5$Å. This length is comparable to a typical bond length in a molecule. This suggests that quantum effects may be important for chemical reactions involving displacement of a hydrogen atom.

The de Broglie wavelength corresponding to the thermal velocity is inversely proportional to the square root of the temperature and to the square root of the mass, $\lambda \propto 1/\sqrt{mT}$. Hence quantum effects are less significant for heavier atoms and at higher temperatures. A quantum treatment of chemical reactions occurring at very low temperatures (e.g., in deep space) may, on the other hand, be imperative.

An electron has a mass that is about 2000 times lower than that of a hydrogen atom, resulting in a thermal de Broglie wavelength of $\lambda \sim 100$Å. Quantum mechanics must be used to describe electrons in most chemical problems! In many (but not all) cases, however, it is possible to avoid explicit treatment

Molecular Kinetics in Condensed Phases: Theory, Simulation, and Analysis,
First Edition. Ron Elber, Dmitrii E. Makarov and Henri Orland.
© 2020 John Wiley & Sons Ltd. Published 2020 by John Wiley & Sons Ltd.

of electronic degrees of freedom by using the so-called Born-Oppenheimer approximation: this approximation takes advantage of the fact that electrons move much faster than nuclei. When nuclei move they experience an effective *average* potential created by the electrons. This potential is a function of the nuclear coordinates only, with the electronic motion averaged out of the problem. Moreover, this effective potential, for any given nuclear configuration, is estimated by assuming that this configuration is fixed. This procedure results in the so-called Born-Oppenheimer potential energy surface (or, simply, potential energy surface, PES), which is a function of the nuclear coordinates. Existence of such a surface was assumed throughout this book. It is, however, important to keep in mind the following two points:

(i) A PES only provides an approximate description of nuclear dynamics regardless of whether or not the nuclei are treated quantum mechanically; in general, coupled dynamics of both electrons and the nuclei must be considered to solve the problem exactly. Moreover, there are important cases (not considered in this book) where the Born-Oppenheimer approximation is inadequate.

(ii) Computing the PES of a molecule from first principles requires solving the quantum problem for the electrons (with fixed nuclei), a task that is often prohibitive computationally, especially for molecules consisting of many atoms. In many applications involving, e.g., large biomolecules or materials, empirical PES's are used instead. Such PES's are further discussed in Chapter 13 of this book.

12.2 How Do the Laws of Quantum Mechanics Affect the Observed Transition Rates?

To illustrate the role of quantum mechanical effects in chemical kinetics, we start with the classical transition state theory expression for the rate of escape from a one-dimensional potential well shown in Fig. 12.1.

This expression is obtained by estimating the flux $q_{A \to B}$ of trajectories crossing the transition state $x = x^{\ddagger}$ (cf. Eqs. (8.9), (8.12), and (8.15), where we identify the reaction coordinate Φ with x):

$$k_{A \to B} = k_{TST} = \frac{q_{A \to B}}{(Z_A/Z)} = Z_A^{-1} \iint_{p>0} \frac{p}{m} \frac{dpdx}{2\pi\hbar} e^{-\beta H(x,p)} \delta(x - x^{\ddagger})$$

$$= Z_A^{-1} \iint_{p>0} \frac{p}{m} \frac{dp}{2\pi\hbar} e^{-\beta\left[\frac{p^2}{2m} + U(x^{\ddagger})\right]}, \tag{12.2}$$

where Z_A is the reactant partition function and $Z = Z_A + Z_B$ the total partition function. Writing the energy of the system in the form

$$E = \frac{p^2}{2m} + U(x^{\ddagger}) \equiv \frac{p^2}{2m} + U^{\ddagger}, \tag{12.3}$$

Figure 12.1 The problem of quantum escape from a potential well (cf. Fig. 10.1). If the energy E of the system is below the barrier energy $U^{\ddagger}$, then the motion in the region $x_1 < x < x_2$, where $U(x_1) = U(x_2) = E$, is classically forbidden.

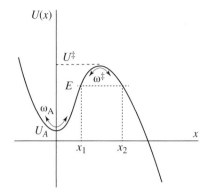

where p is the momentum at the moment the barrier is crossed, we have

$$pdp/m = dE, \qquad (12.4)$$

which allows us to rewrite Eq. (12.2) as an average over energy:

$$k_{TST} = Z_A^{-1} \int_{U_A}^{\infty} \frac{dE}{2\pi\hbar} \theta[E - U^{\ddagger}]e^{-\beta E}, \qquad (12.5)$$

The Heaviside theta function restricts the integration over energy to the classically allowed energies that are higher than the barrier, $E > U^{\ddagger}$.

To account for quantum effects, two heuristic extensions of Eq. (12.5) may be suggested. First, instead of using the classical expression for the reactant partition function Z_A as before, the quantum one can be used. Assuming the harmonic approximation in at the bottom of the well corresponding to the state A and using the textbook expression for the partition function of a harmonic oscillator [1] we write

$$Z_A \approx \frac{e^{-\beta U_A}}{2\sinh(\beta\hbar\omega_A/2)}, \qquad (12.6)$$

where ω_A is the vibrational frequency in the reactant well (Figure 12.1). Second, quantum mechanics allows classically forbidden transitions: for example, a particle with an energy E that is below the barrier may still end up on the product side of the barrier via the quantum tunneling effect. In the classical expression for the rate, Eq. (12.5), the Heaviside theta function precludes such nonclassical transitions, as it disallows energies such that $E < U^{\ddagger}$. We can relax the classical requirement for the energy to be above $U^{\ddagger}$ and allow the possibility of crossing the barrier even when the energy is below the barrier $U^{\ddagger}$ by replacing $\theta[E - U^{\ddagger}]$ in Eq. (12.5) with a quantum mechanical "transmission coefficient" (or "barrier transparency") $Tr(E)$: this coefficient accounts for the probability of penetrating the barrier via the tunneling process. The proposed quantum expression for the quantum rate thus becomes:

$$k_{A \to B} = k_q = Z_A^{-1} \int_{-\infty}^{\infty} \frac{dE}{2\pi\hbar} Tr(E)e^{-\beta E} \qquad (12.7)$$

For $E \gg U^{\ddagger}$, we expect the transmission coefficient to approach one. Interestingly, for an energy E that is not too far above the barrier $U^{\ddagger}$, the transmission coefficient is not equal to one: Quantum mechanics also allows for the possibility of reflection from the barrier even when the energy of the system exceeds that of the barrier. The tunneling probability decreases as the energy becomes lower. We thus expect $Tr(E)$ to resemble the "classical transmission factor $\theta[E - U^{\ddagger}]$" in shape but to be a continuous rather than a stepwise function. The derivation of the expression for $Tr(E)$ requires solving she Schrödinger equation for the potential $U(x)$ and is beyond the scope of this book. Instead, we refer the interested reader to quantum mechanics textbooks, such as the one by Landau and Lifshitz [2], for the derivation of the expressions that will be used before.

If, as before, we approximate the potential near the barrier top by an inverted parabola, $U(x) \approx U^{\ddagger} - m(\omega^{\ddagger})^2(x - x^{\ddagger})^2/2$ (Fig. 12.1), then $Tr(E)$ can be calculated exactly [2]:

$$Tr(E) = \frac{1}{1 + \exp\left[\frac{2\pi}{\hbar\omega^{\ddagger}}(U^{\ddagger} - E)\right]} \tag{12.8}$$

Notice that this function approaches unity at high energies and zero at low energies, just like the Heaviside theta. Moreover, the transition between these two limits becomes sharp approaching the Heaviside function in the limit $\hbar\omega^{\ddagger} \to 0$ - a classical limit is reached in this case.

Substituting Eq. (12.8) into Eq. (12.7) and evaluating the integral, one obtains

$$k_q = \frac{\hbar\omega^{\ddagger}\beta/2}{\sin(\hbar\omega^{\ddagger}\beta/2)} \frac{1}{2\pi\hbar\beta Z_A} e^{-\beta U^{\ddagger}} = \frac{\hbar\omega^{\ddagger}\beta/2}{\sin(\hbar\omega^{\ddagger}\beta/2)} k_{TST} \tag{12.9}$$

Quantum mechanical effects enter into Eq. (12.9) in two distinct ways. First, the quantum partition function (Eq. (12.6)) instead of the classical one is used to evaluate k_{TST}. In the low temperature limit, $\beta\hbar\omega_A \gg 1$, we can approximate Eq. (12.6) by $Z_A \approx \exp[-\beta(U_A + \hbar\omega_A/2)]$. As a result, the rate predicted by Eq. (12.9) is proportional to the factor $Z_A \approx \exp[-\beta(U^{\ddagger} - U_A - \hbar\omega_A/2)]$. This result has a simple physical interpretation: at low temperature quantum effects raise the initial energy of the system (in state A) from the bottom of the potential well ($U = U_A$) to $U = U_A + \hbar\omega_A/2$, where $\hbar\omega_A/2$ is the zero-point vibrational energy in the harmonic well. As a result, the total activation barrier, $U^{\ddagger} - U_A - \hbar\omega_A/2$, becomes lower than that in the classical case. Although intuitively appealing, this physical argument is too simplistic because, as will be seen shortly, Eq. (12.9) breaks down quite dramatically in the lower temperature case.

Second, according to Eq. (12.9), quantum corrections have further amplified the classical transition state theory rate expression k_{TST} by a factor equal to

$$c = \frac{\hbar\omega^{\ddagger}\beta/2}{\sin(\hbar\omega^{\ddagger}\beta/2)} \tag{12.10}$$

At high temperatures, $\hbar\omega^{\ddagger}\beta \ll 1$, we have $\sin(\hbar\omega^{\ddagger}\beta/2) \approx \hbar\omega^{\ddagger}\beta/2$, which gives $c \approx 1$ - the transition state theory is recovered. As the temperature is lowered, however, the value of c is increased, until it becomes infinite at $\beta\hbar\omega^{\ddagger}/2 = \pi$. This is obviously an unphysical result, and something must be wrong with our theory when the temperature becomes too low. Before examining the physical reasons for this divergence, we first define the so-called *crossover temperature* as the temperature at which c diverges:

$$T_c = \frac{1}{k_B\beta_c} = \frac{\hbar\omega^{\ddagger}}{2\pi k_B} \tag{12.11}$$

To understand the physical cause of this divergence, let us examine the integral of Eq. (12.7) more closely. It contains a product of two functions. The Boltzmann factor $e^{-\beta E}$ increases with decreasing energy E reflecting that the low-lying initial states are increasingly more populated. In contrast, the transmission factor $Tr(E)$ decreases with decreasing energy reflecting that penetrating the barrier via tunneling becomes increasingly improbable. For $U^{\ddagger} - E \gg \hbar\omega^{\ddagger}/(2\pi)$, the behavior of the transmission factor (Eq.(12.8)) can be approximated by an exponential function,

$$Tr(E) \approx \exp\left[\frac{2\pi}{\hbar\omega^{\ddagger}}(E - U^{\ddagger})\right] = \exp\left[\frac{E - U^{\ddagger}}{k_B T_c}\right] \tag{12.12}$$

When $T > T_c$, the Boltzmann factor increases slowly enough that the product $e^{-\beta E} Tr(E)$ vanishes in the limit $E \to -\infty$. The integral in Eq. (12.7) then converges. In contrast, for $T < T_c$ the integral diverges, as the states with low energy E dominate. This divergence is readily eliminated if we recognize that the lower integration limit in Eq. (12.7) should be not minus infinity but the minimum possible value of the energy in the potential well, which cannot be lower than U_A. Furthermore, the parabolic barrier approximation should not be used in this case, as it is obviously quite poor near the bottom of the well. Equation (12.9) is a good approximation only when the dominant contribution in the integral of Eq. (12.7) comes from energies that are not too far from $U^{\ddagger}$ where the parabolic barrier approximation to the transmission factor, Eq. (12.8), is valid, and this is not true for any temperature below T_c. Hence the low temperature case, $T < T_c$, requires a special treatment that takes the true shape of the potential $U(x)$ into consideration.

12.3 Semiclassical Approximation and the Deep Tunneling Regime

The parabolic barrier approximation, Eq. (12.8), is inadequate at low temperatures where energies close to the bottom of the potential well dominate the integral of Eq. (12.7). The correct expression for the transmission factor $Tr(E)$

then depends on the precise shape of the potential and, in general, cannot be written in analytical form. A useful approximation for $Tr(E)$ can however be obtained in the so-called semiclassical limit. The semiclassical limit of quantum mechanics is a very important one. For example, the Bohr model of the hydrogen atom is an example of semiclassical theory. The following heuristic arguments constitute a rather crude and sloppy explanation of the physical basis of the semiclassical approximation; again, the reader interested in a deeper understanding of the subject is referred to quantum mechanics textbooks [2].

Let us start with the textbook problem of a quantum mechanical particle of mass m and energy E subjected to the step-wise potential of the form (Fig. 12.2):

$$U(x) = \begin{cases} 0, x < 0 \\ U_0, x \geq 0 \end{cases} \tag{12.13}$$

The wavefunction of the particle, $\psi(x)$, satisfies the Schrödinger equation,

$$-\frac{\hbar^2}{2m}\psi'' + U(x)\psi = E\psi \tag{12.14}$$

We are particularly interested in the solution of this equation for energies E below the barrier, $E < U_0$, as those describe tunneling below the barrier. For $x>0$ we obtain

$$\psi'' = \frac{2m(U_0 - E)}{\hbar^2}\psi, \tag{12.15}$$

and the solution of Eq. (12.15) is of the general form

$$\psi(x) = a \exp\left[-\frac{\sqrt{2m(U_0 - E)}}{\hbar}x\right] + b \exp\left[\frac{\sqrt{2m(U_0 - E)}}{\hbar}x\right], \tag{12.16}$$

where a and b are some coefficients. This wavefunction is a sum of two terms, one of which decaying exponentially with x and the other growing exponentially. Since a physically meaningful wavefunction cannot grow without bounds, the exponentially growing term has to be discarded as unphysical. Keeping

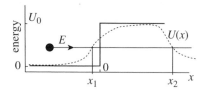

Figure 12.2 Quantum motion of a particle with an energy E in a potential barrier. The motion in the region where $U(x) > E$ is forbidden classically. A quantum particle approaching the barrier from the left will, however, be found in the classically forbidden region with a nonzero probability.

only the first term in Eq. (12.16), one concludes that the probability density of finding the system at x decays exponentially,

$$p(x) = |\psi(x)|^2 \sim \exp\left[-2\frac{\sqrt{2m(U_0 - E)}}{\hbar}x\right] \tag{12.17}$$

A crude physical interpretation of Eq. (12.17) is that, if a particle with an energy $E < U_0$ approaches the barrier from the left, it may "penetrate" the barrier via tunneling and be found a distance x away from the barrier entrance with a probability that decays exponentially with this distance.

Consider now the case where, instead of a barrier with a constant potential, we have a potential $U(x)$ that is varied in space. If this spatial variation is sufficiently slow, we can replace Eq. (12.17) with an approximate formula

$$p(x) = |\psi(x)|^2 \sim \exp\left\{-2\int_{x_1}^{x} dx'\frac{\sqrt{2m[U(x') - E]}}{\hbar}\right\}, \tag{12.18}$$

where x_1 is the location of the classical turning point where $U(x_1) = E$ such that $E > U(x)$ for $x < x_1$ (Figure 12.2). For a barrier of finite width, as illustrated in Fig. 12.2, the probability that the system will emerge on the other side will be

$$p(x_2) \sim \exp\left\{-2\int_{x_1}^{x_2} dx\frac{\sqrt{2m[U(x) - E]}}{\hbar}\right\}. \tag{12.19}$$

Here x_2 is the other turning point where, again, $U(x_2) = E$, so that the motion in the interval $x_1 < x < x_2$ is classically forbidden. It is then plausible that the transmission coefficient can be estimated as

$$Tr(E) \sim e^{-2S(E)/\hbar}, \tag{12.20}$$

where we introduced the "action"

$$S(E) = \int_{x_1}^{x_2} dx\sqrt{2m[U(x) - E]} \tag{12.21}$$

Adopting the approximation given by Eq. (12.20) (again, its proper justification and a discussion of when it is accurate is beyond the scope of this Chapter), we rewrite the integral of Eq. (12.7) as

$$k_{A \to B} = k_{A \to B}^q = Z_A^{-1}\int_{U_A}^{\infty}\frac{dE}{2\pi\hbar}e^{-\beta E - 2S(E)/\hbar} \tag{12.22}$$

The dominant contribution into this integral arises from energies in the vicinity $E \approx E_m$ of the minimum of the function

$$\beta E + 2S(E)/\hbar, \tag{12.23}$$

which can be found by setting the derivative of Eq. (12. 23) with respect to the energy E to zero:

$$\beta + 2S'(E_m)/\hbar = 0 \tag{12.24}$$

The result can be written in a physically appealing form by using the definition of action (Eq. (12.21)) and then noting that

$$S'(E) = -\int_{x_1}^{x_2} dx \sqrt{\frac{m}{2[U(x) - E]}} \equiv -\int_{x_1}^{x_2} \frac{dx}{v(x)}, \tag{12.25}$$

where the "velocity" $v(x)$ satisfies the equation

$$\frac{mv^2(x)}{2} = -[E - U(x)]. \tag{12.26}$$

Equation (12.26) is a statement of conservation of energy in the upside-down potential $-U(x)$. Therefore, the derivative of the action with respect to energy in Eq. 12.25 is equal to the time that a particle moving in the upside down potential takes to traverse the barrier from the turning point x_1 to the turning point x_2, taken with the negative sign. Introducing the oscillation period in the upside-down potential, $\tau(E)$, we have

$$S'(E) = -\tau(E)/2 \tag{12.27}$$

Eq. 12.24 can then be restated as

$$\tau(E_m) = \beta\hbar, \tag{12.28}$$

We can now evaluate the integral of Eq. (12.22) by approximating the action, in vicinity of E_m, by its Taylor series truncated at second order,

$$S(E) \approx S(E_m) + S'(E_m)(E - E_m) + \frac{1}{2}S''(E_m)(E - E_m)^2, \tag{12.29}$$

and extending the integration limits to infinity. This gives

$$k_q \approx \frac{1}{2\pi\hbar Z_A} \sqrt{\frac{\pi\hbar}{|S''(E_m)|}} e^{-\beta E_m - 2S(E_m)/\hbar} = \frac{1}{\sqrt{2\pi\hbar|\tau'(E_m)|}Z_A} e^{-\beta E_m - 2S(E_m)/\hbar} \tag{12.30}$$

Let us now examine the behavior of the semiclassical result, Eq. (12.30), as a function of temperature. If the curvature of the potential at the barrier top is nonzero (i.e. $\omega^{\ddagger} \neq 0$), then the minimum possible value of the oscillation period in the upside-down potential is

$$\min_E \tau(E) = \frac{2\pi}{\omega^{\ddagger}}. \tag{12.31}$$

This value corresponds to oscillations with a vanishingly small amplitude. An oscillatory solution to Eq. 12.28 does not exist for $\beta\hbar < 2\pi/\omega^{\ddagger}$, or, equivalently, for $T > T_c$, where the crossover temperature T_c is defined by Eq. (12.11). In this case the only possibility to satisfy Eq. (12.28) is to consider a trajectory that shrank into a point, and the system is permanently situated at the barrier top, $x = x^{\ddagger}$. For such a trajectory we have $E_m = U^{\ddagger}$, $S(E_m) = 0$. Although we should not be using the semiclassical approximation in this case, it is reassuring to see that Eq. (12.30) predicts an Arrhenius law with a rate proportional to $e^{-\beta U^{\ddagger}}$, albeit with a nonsensical prefactor. This is a regime where Eq. (12.9) could be used instead to estimate the quantum corrections to the classical rate.

Below the crossover temperature, Eq. (12.28) can now be satisfied by an oscillatory trajectory in the upside-down potential, whose oscillation period matches $\beta\hbar$. As the oscillation period becomes longer, the amplitude becomes larger and, correspondingly, the temperature becomes lower (i.e. $\beta\hbar$ becomes larger). Consider now the limit where the temperature approaches zero, $T \to 0$. This corresponds to a periodic trajectory whose period approaches infinity. To understand this limit better, consider motion in the upside-down potential, as shown in Figure 12.3. Without loss of generality, we can set the energy of the A-well minimum (which now becomes a maximum) to zero, $U_A = 0$. Likewise, we can set the position of the minimum (upside down potential maximum) to be $x_A = 0$. If we place the system at a very small distance $x_0 > 0$ to the right of this point with zero initial velocity ($\dot{x}(0) = 0$), it will slowly slide toward the right, undergo an oscillation, and then climb back toward the point x_0. The limit $\tau \to \infty$ is achieved when $x_0 \to 0$ and, accordingly, $E = m\omega_A^2 x_0^2/2 \to 0$. For a very small x_0, then, most of the oscillation period τ will be spent in the vicinity of the point $x = x_0$.

We can make the above statement more quantitative as follows. Pick some intermediate point $x_1 > 0$ that is close enough to $x = 0$ such that the harmonic approximation for the potential $U(x_1) \approx m\omega_A^2 x_1^2/2$ holds to the desired accuracy. We can then write

$$\tau(E) = \tau(m\omega_A^2 x_0^2/2) = 2\tau(x_0 \to x_1) + \tau(x_1 \rightleftarrows x_1) \tag{12.32}$$

The first term in the r.h.s. of Eq. (12.32) is twice the time for the system to slide x_0 from to x_1, and the second term is the roundtrip time starting from x_1, going right and back. In the limit $x_0 \to 0$, $\tau \to \infty$, the first term diverges while the second term does not, as the maximum possible value for $\tau(x_1 \rightleftarrows x_1)$, which is achieved when the velocity at x_1 is zero (i.e. when $x_1 = x_0$), is finite. Thus, for sufficiently large values of $\tau = \beta\hbar$ we can neglect the second term and write

$$\tau(E) \approx 2\tau(x_0 \to x_1) \tag{12.33}$$

Figure 12.3 Motion in the upside down potential $-U(x)$: a periodic orbit starts, with zero velocity, at $x = x_0$, and passes an intermediate point $x = x_1$. Harmonic approximation is used to describe the potential in the region $x_0 \leq x \leq x_1$.

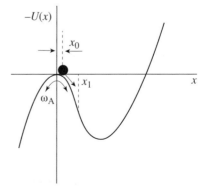

Since the harmonic approximation can be used within the interval $x \le x_1$, the part of the oscillating trajectory confined to this interval obeys the Newtonian dynamics in the potential $-U$:

$$m\ddot{x} = U'(x) = m\omega_A^2 x, \tag{12.34}$$

and the solution satisfying the initial condition of zero velocity at $t=0$ is

$$x(t) = x_0 \cosh \omega_A t \tag{12.35}$$

We thus have

$$x_1 = \cosh[\omega_A \tau(x_0 \to x_1)] = \frac{1}{2} \exp[\omega_A \tau(x_0 \to x_1)]$$
$$+ \frac{1}{2} \exp[-\omega_A \tau(x_0 \to x_1)]$$
$$\approx \frac{1}{2} \exp[\omega_A \tau(x_0 \to x_1)], \tag{12.36}$$

where one of the two exponential terms could be dropped as it vanishes in the limit $\tau \to \infty$ considered here. We now can use Eqs.12.33 and 12.36 to obtain

$$\tau(E) \approx \frac{2}{\omega_A} \ln \frac{2x_1}{x_0} = \frac{1}{\omega_A} \ln \frac{4\omega_A^2 x_1^2}{\omega_A^2 x_0^2} = \frac{1}{\omega_A} \ln \frac{4U(x_1)}{E}, \tag{12.37}$$

from where it follows that

$$\frac{d\tau}{dE} = -\frac{1}{\omega_A E}, \tag{12.38}$$

or

$$E = E_0 \exp(-\omega_A \tau), \tag{12.39}$$

where E_0 is a constant that is determined by the shape of the potential $U(x)$ and that is of the same order of magnitude as the height of the barrier. This result can now be used to find the zero-temperature limit of the semiclassical equation, Eq. (12.30): Using the harmonic approximation for the reactant's partition function,

$$Z_A = \frac{1}{2 \sinh \beta \hbar \omega_A / 2} \approx e^{-\beta \hbar \omega_A / 2} = e^{-\omega_A \tau / 2} \tag{12.40}$$

we see that the temperature disappears from the final answer:

$$k_q(T \to 0) \approx \frac{\omega_A}{2\pi} \sqrt{\frac{2\pi E_0}{\hbar \omega_A}} e^{-2S(0)/\hbar}, \tag{12.41}$$

where

$$S(0) = \int_0^{x^{\ddagger}} dx \sqrt{2mU(x)} \tag{12.42}$$

Equation (12.41) resembles the result of the classical transition state theory, Eq. (8.28), with the Boltzmann exponent $\beta U^{\ddagger}$ replaced by the temperature independent quantity $2S(0)/\hbar$. There is an additional factor $\sqrt{2\pi E_0/\hbar\omega_A}$, which is normally greater than 1. It can be shown that this factor accounts for the fact that the lowest possible energy in the potential well is not zero but the zero-point vibrational energy $\hbar\omega_A/2$; as a result, the barrier transmission factor $Tr(E)$ should be evaluated not at $E=0$ as in Eq. (12.42) but at the zero-point energy, leading to a higher value of the rate.

Combining our results above the crossover temperature, where the transition rate exhibits Arrhenius behavior, with the quantum theory below crossover, where a quantum, tunneling-dominated low-temperature plateau is reached at low temperatures, one obtains a dependence $k_{A \to B}(T)$ that is schematically depicted in Figure 12.4. It is convenient to represent this dependence in the form of an "Arrhenius plot", with the logarithm of the rate coefficient plotted as a function of the inverse temperature, $1/T$ (Fig. 12.4). In the high-temperature, Arrhenius law regime, then, the log of the rate decreases (approximately) linearly with increasing $1/T$, but it eventually levels off reaching the quantum limit. The transition between the classical and the quantum regimes takes place in the vicinity of the crossover temperature T_c.

The "deep tunneling" regime with a temperature independent transition rate has been observed for an number of chemical reactions at very low temperatures, typically tens of degrees Kelvin [3]. Such chemical processes may be important, for example, in the interstellar medium. Tunneling effects are usually more subtle under less exotic conditions, such as at room temperature. As noted in the beginning of this Chapter, quantum effects are most pronounced for light particles, particularly for protons. Indeed, the crossover temperature, Eq. (12.11), is proportional to the oscillation frequency $\omega^{\ddagger}$ and thus inversely proportional to the square root of the mass. Thus, tunneling effects are manifested at higher temperature for lighter particles. In particular, the crossover temperature for reactions involving transfer of proton is often comparable to room temperature [3]. Many biochemical transformations in living organisms involve such proton transfer reactions, and quantum effects need be considered

Figure 12.4 "Arrhenius plot" of the quantum transition rate as a function of temperature.

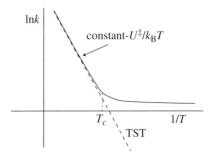

when studying those processes. Yet quantum low-temperature plateau is rarely reached, and the quantum corrections provided by Eq. (12.9) often provide a sufficient account of the quantum effects.

12.4 Path Integrals, Ring-Polymer Quantum Transition-State Theory, Instantons and Centroids

The semiclassical theory described above provides a solution to the one-dimensional model of a chemical reaction involving tunneling along a single degree of freedom. This model is rarely realistic, since most chemical processes involve coupled motions of many atoms. It turns out that a multidimensional generalization is not easy to accomplish when tunneling effects are important.

Friction or energy dissipation effects present a further difficulty for quantum rate theories. In the classical case such effects could be captured by introducing non-conservative friction forces, leading to the Kramers, Grote-Hynes, and Langer theories discussed in the preceding Chapters. The fundamental equations describing quantum mechanical time evolution, however, do not allow for non-conservative forces. The solution of this conceptual problem lies in the idea that was used in Chapter 9: one recognizes that friction is a result of projecting conservative dynamics of a much larger system onto a low-dimensional space containing the degrees of freedom of interest, as illustrated by our derivation of the Langevin equation from the Zwanzig-Caldeira-Leggett Hamiltonian. By considering all degrees of freedom explicitly, the conceptual challenge is solved, but a practical challenge arises: one must be able to describe quantum effects in a system of high dimensionality.

Many methods invented to describe quantum transitions in high-dimensional molecular systems take advantage of Feynman's path integral formulation of quantum mechanics. Some of these ideas are quite elegant, and the analogies that arise between classical and quantum systems in this formulation are thought-provoking. The following notes are not a systematic introduction to the subject but merely an attempt to provide a glimpse into this fascinating subject. A brief derivation of the path integral formalism is given, assuming that the reader is familiar with the basics of quantum mechanics such as the time-dependent and time-independent Schrödinger equations and with Dirac's notation.

The time-independent Schrödinger equation is the eigenvalue problem for the system's Hamiltonian operator $\hat{H}$. It determines the possible energy levels E_n and the corresponding stationary states $|n\rangle$ of the system:

$$\hat{H}|n\rangle = E_n|n\rangle. \tag{12.43}$$

For a one-dimensional particle of mass m moving in a potential $U(x)$, the quantum Hamiltonian is given by

$$\hat{H} = \frac{\hat{p}^2}{2m} + \hat{U}, \tag{12.44}$$

where $\hat{p} = (\hbar/i)d/dx$ is the momentum operator and $\hat{U}$ is the potential energy operator, whose action on an arbitrary function $\psi(x)$ is defined by

$$\hat{U}\psi(x) = U(x)\psi(x) \tag{12.45}$$

The operator $\hat{U}$ is diagonal in the position representation,

$$\hat{U}|x_1\rangle = U(x_1)|x_1\rangle, \tag{12.45a}$$

where $|x_1\rangle$ describes the state whose wavefunction is the delta function, $\psi(x) = \delta(x - x_1)$.

The time evolution of the system's state $|\psi(t)\rangle$ is determined by the time-dependent Schrödinger equation

$$i\hbar \frac{\partial}{\partial t}|\psi(t)\rangle = \hat{H}|\psi(t)\rangle, \tag{12.46}$$

whose solution is

$$|\psi(t)\rangle = e^{-\frac{i}{\hbar}\hat{H}t}|\psi(0)\rangle, \tag{12.47}$$

so that $e^{-\frac{i}{\hbar}\hat{H}t}$ can be viewed as the time evolution operator. In the coordinate representation, Eq. (12.47) can be rewritten as an equation describing the time evolution of a wavefunction, $\psi(x, t) \equiv \langle x|\psi(t)\rangle$:

$$\psi(x, t) = \int dx_0 \langle x|e^{-\frac{i}{\hbar}\hat{H}t}|x_0\rangle \psi(x_0, 0) \tag{12.48}$$

The path integral formalism provides a tantalizing formula for $\langle x|e^{-\frac{i}{\hbar}\hat{H}t}|x_0\rangle$, a quantity that can be thought of as the "probability amplitude" that the system starting in x_0 at $t=0$ will reach x at time t. To derive this formula, let us first consider the behavior of $\langle x|e^{-\frac{i}{\hbar}\hat{H}\Delta t}|x_0\rangle$ at short times, $\Delta t \to 0$. We write:

$$\langle x|e^{-\frac{i}{\hbar}\hat{H}\Delta t}|x_0\rangle \approx \langle x|e^{-\frac{i}{\hbar}\frac{\hat{p}^2}{2m}\Delta t}e^{-\frac{i}{\hbar}\hat{U}\Delta t}|x_0\rangle = \langle x|e^{-\frac{i}{\hbar}\frac{\hat{p}^2}{2m}\Delta t}|x_0\rangle e^{-\frac{i}{\hbar}U(x_0)\Delta t} \tag{12.49}$$

Eq. (12.49) deserves an explanation. First, we have used the identity

$$e^{-\frac{i}{\hbar}\hat{U}\Delta t}|x_0\rangle = \sum_{n=0}^{\infty} \frac{1}{n!}\left(-\frac{i}{\hbar}\hat{U}\Delta t\right)^n |x_0\rangle$$

$$= \sum_{n=0}^{\infty} \frac{1}{n!}\left(-\frac{i}{\hbar}U(x_0)\Delta t\right)^n |x_0\rangle = e^{-\frac{i}{\hbar}U(x_0)\Delta t}|x_0\rangle,$$

which follows from Eq. (12.45a). Second, we have used the so-called Trotter formula,

$$e^{\hat{A}\Delta t + \hat{B}\Delta t} \approx e^{\hat{A}\Delta t}e^{\hat{B}\Delta t}$$

where $\hat{A} = (-i/\hbar)\hat{p}^2/(2m)$ and $\hat{B} = (-i/\hbar)\hat{U}$. In general, this is an approximation for any finite Δt, unless the operators $\hat{A}$ and $\hat{B}$ commute; it becomes, however, exact in the limit of $\Delta t \to 0$.

To get rid of the kinetic energy operator in Eq. (12.49), it is expedient to use the momentum representation involving states $|p_1\rangle$ such that $\hat{p}|p_1\rangle = p_1|p_1\rangle$. It can be easily verified that these states are plane waves in the position representation: i.e., the corresponding wavefunctions are given by $\psi(x) = \langle x|p_1\rangle = \frac{1}{\sqrt{2\pi\hbar}}e^{ip_1 x/\hbar}$. Using the identity $\int dp_1 |p_1\rangle\langle p_1| = 1$, we have

$$\langle x|e^{-\frac{i}{\hbar}\frac{\hat{p}^2}{2m}\Delta t}|x_0\rangle = \int dp_1 \int dp_2 \langle x|p_1\rangle\langle p_1|e^{-\frac{i}{\hbar}\frac{\hat{p}^2}{2m}\Delta t}|p_2\rangle\langle p_2|x_0\rangle$$

$$= \int dp_1 \langle x|p_1\rangle e^{-\frac{i}{\hbar}\frac{p_1^2}{2m}\Delta t}\langle p_1|x_0\rangle$$

$$= \frac{1}{2\pi\hbar}\int dp_1 e^{-\frac{i}{\hbar}\frac{p_1^2}{2m}\Delta t + \frac{i}{\hbar}p_1(x-x_0)} = \left(\frac{m}{2\pi\hbar i\Delta t}\right)^{1/2} e^{i\frac{m}{2\hbar}\left(\frac{x-x_0}{\Delta t}\right)^2 \Delta t}$$

$$(12.50)$$

Eq. 12.49 can, therefore, be rewritten as

$$\langle x|e^{-\frac{i}{\hbar}\hat{H}\Delta t}|x_0\rangle \approx \left(\frac{m}{2\pi\hbar i\Delta t}\right)^{1/2} e^{i\frac{m}{2\hbar}\left(\frac{x-x_0}{\Delta t}\right)^2 \Delta t} e^{-\frac{i}{\hbar}U(x_0)\Delta t} \tag{12.51}$$

For a finite time interval t, we divide it into $N \gg 1$ time slices, $t = N\Delta t$, and write

$$\langle x|e^{-\frac{i}{\hbar}\hat{H}t}|x_0\rangle = \langle x|\left(e^{-\frac{i}{\hbar}\hat{H}\Delta t}\right)^N|x_0\rangle$$

$$\approx \int dx_1\, dx_2 \ldots dx_{N-1}\langle x|e^{-\frac{i}{\hbar}\hat{H}\Delta t}|x_{N-1}\rangle$$

$$\times \langle x_{N-1}|e^{-\frac{i}{\hbar}\hat{H}\Delta t}|x_{N-2}\rangle \ldots \langle x_1|e^{-\frac{i}{\hbar}\hat{H}\Delta t}|x_0\rangle$$

$$= \left(\frac{m}{2\pi\hbar i\Delta t}\right)^{N/2}\int dx_1 dx_2 \ldots dx_{N-1}e^{iS_L(x_0,x_1,\ldots,x)/\hbar} \tag{12.52}$$

where

$$S_L(x_0, x_1, \ldots, x_N) = \Delta t \sum_{n=1}^{N}\left[\frac{m(x_n - x_{n-1})^2}{\Delta t^2} - U(x_{n-1})\right] \tag{12.53}$$

We can think of the sequence $x_0, x_1, \ldots, x_N = x$ as a discretized version of a trajectory $x(t')$, where $x_n = x(n\Delta t)$. The trajectory starts at $x(0) = x_0$ and ends at $x(t) = x$. In the limit $N \to \infty$, then, Eq. (12.53) becomes the classical Lagrangian action of the particle:

$$S_L[x(t')] = \int_0^t dt'\left\{\frac{m\dot{x}^2}{2} - U[x(t')]\right\} \tag{12.54}$$

We further introduce integration over all possible paths $x(t')$, i.e. integration over all possible intermediate points in the limit where the number of such points becomes infinite:

$$\lim_{N\to\infty}\left(\frac{m}{2\pi\hbar i\Delta t}\right)^{N/2}\int dx_1 dx_2 \ldots dx_{N-1} \ldots \equiv D[x(t')] \tag{12.55}$$

Using this notation, we arrive at Feynman's famous formula

$$\langle x|e^{-\frac{i}{\hbar}\hat{H}t}|x_0\rangle = \int_{x(0)=x_0}^{x(t)=x} D[x(t')]e^{iS_L[x(t')]/\hbar} \tag{12.56}$$

Crudely speaking, if we want to compute the amplitude to go from x_0 to x over a time interval t, we must consider all possible paths $x(t')$ connecting these points, each of which will contribute $e^{iS_L[x(t')]/\hbar}$ to the total amplitude.

One of the attractive features of Feynman's formulation of quantum mechanics is that it naturally "explains" how classical mechanics arises as a limit of quantum mechanics. Specifically, when the typical value of the Lagrangian action S_L far exceeds Planck's constant, the complex exponential $e^{iS_L[x(t')]/\hbar}$ will oscillate wildly, and contributions from most of the paths will cancel each other out. The only exceptions are the paths in the vicinity of which the action changes very slowly. It is, perhaps, simpler to understand this idea using regular integration as an example. Consider an integral of the form $\int dx e^{i\Omega(x)/\hbar}$, where $\Omega(x)$ is a real-valued function such that $|\Omega(x)| \gg \hbar$. Because the complex exponential oscillates wildly, the dominant contribution to the integral comes from the points x_{st} where $\Omega(x)$ is stationary, i.e., where $\Omega'(x_{st}) = 0$. Those are the maxima or minima of the function $\Omega(x)$. By analogy, the dominant paths emerging in the classical limit ($\hbar \to 0$) are the extrema of the Lagrangian action, Eq. (12.54). It is well known in classical mechanics that those are the classical trajectories satisfying Newton's equations of motion,

$$m\ddot{x} = -U'(x) \tag{12.57}$$

The path integral formulation can also be applied to describe quantum statistical mechanics. To this end, consider the quantum partition function for a system satisfying Eq. (12.43),

$$Z = \sum_n e^{-\beta E_n} = \sum_n \langle n|e^{-\beta\hat{H}}|n\rangle = \mathrm{tr}\, e^{-\beta\hat{H}}, \tag{12.58}$$

where "tr" stands for the trace of an operator. Because trace is invariant with respect to the representation, it can be equivalently expressed using the position representation,

$$Z = \int dx\langle x|e^{-\beta\hat{H}}|x\rangle \tag{12.59}$$

Now, the quantity $\langle x|e^{-\beta \hat{H}}|x\rangle$ can be found from Eq. (12.56) if we formally make the substitution

$$\beta = it/\hbar \qquad (12.60)$$

This leads us to consider classical trajectories evolving in imaginary time, $\tau = it'$, and we obtain:

$$Z = \int_{x(0)=x(\beta\hbar)} D[x(\tau)]e^{-S_E[x(\tau)]/\hbar}, \qquad (12.61)$$

where

$$S_E[x(\tau)] = \int_0^{\beta\hbar} d\tau \left\{ \frac{m\dot{x}^2}{2} + U[x(\tau)] \right\} \qquad (12.62)$$

is the Largangian action in the upside-down potential, $-U(x)$! In the literature, this action is usually called the Euclidian action, which is reflected in the subscript. Notice that the complex exponentials are gone: Eq. (12.61) is an integral of a real-valued function.

Equations (12.61–12.62) provide a starting point for several approaches that extend the ideas of classical TST to the realm of quantum mechanics. To explain them, it is expedient to temporarily switch to a discretized version of Eq. (12.62). Dividing the interval $\beta\hbar$ into N time slices with $\Delta\tau = \beta\hbar/N$, we write, similarly to Eq. (12.53),

$$S_E(x_0, x_1, \ldots, x_N) \approx \Delta\tau \sum_{n=1}^N \left[\frac{m(x_n - x_{n-1})^2}{\Delta\tau^2} + U(x_{n-1}) \right] \qquad (12.63)$$

Because integration in Eq. (12.61) involves only closed paths, $x(0) = x(\beta\hbar)$, we have $x_0 = x_N$ in the discretized version of our theory. Eq. (12.63) then suggests a "classical" interpretation: we envision a collection of particles with coordinates $x_0, x_1, \ldots, x_{N-1}, x_0$ forming a ring (Figure 12.5). Each particle is subjected to the external potential $U(x)$; in addition, each pair of adjacent particles is connected

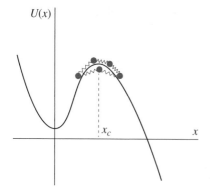

Figure 12.5 In path-integral based transition state theories discussed here, the problem of quantum escape of a particle through a barrier is mapped onto that of classical escape of a polymer ring (formed by multiple replicas of the particle) through the same barrier.

by a linear spring with a spring constant $K = m/\Delta\tau^2$. Then to within a factor, Eq. (12.63) describes the energy of such a polymer ring; a discrete version of Eq. (12.61),

$$Z = \left(\frac{m}{2\pi\hbar\Delta\tau}\right)^{N/2} \int dx_0 \, dx_1 \, dx_2 \dots dx_{N-1} e^{-S_E(x_0,x_1,\dots,x_{N-1},x_0)/\hbar}, \qquad (12.64)$$

would then describe the classical partition function of this ring. We have mapped the quantum statistical mechanics of a single particle in a potential $U(x)$ onto classical statistical mechanics of polymer rings [4]! This result can be straightforwardly generalized to multidimensional systems, where the potential depends on multiple variables.

The transition between the classical and quantum regimes is particularly clear in this picture of quantum statistical mechanics. As temperature is increased, the springs connecting the particles in the ring become stiffer; in the limit $\beta \to 0$ the ring collapses into a single point x, and the quantity $S_E/\hbar$ approaches the Boltzmann exponent $\beta U(x)$ - classical statistical mechanics is recovered.

Given this mapping, it is tempting to identify the quantum transition rate with the *classical* rate for the associated polymer ring to traverse the same barrier, but can this be justified? Eq. (12.64) describes equilibrium properties of the system. By applying a classical simulation method (e.g., molecular dynamics, Monte Carlo etc.) to an extended polymer ring system, any *equilibrium* property of the corresponding quantum system can be rigorously computed. Unfortunately, this idea cannot be rigorously extended to *dynamics*: we cannot (rigorously) claim that classical dynamics of polymer rings is equivalent to quantum dynamics of their quantum counterparts. It turns out that somewhat more rigorous arguments connecting the dynamics of the two systems can be made within the more limiting scope of computing the quantum transition rate between states separated by a sufficiently high barrier [3, 5–8].

One of the most appealing features of Eq. (12.64) is that it provides an opportunity to take advantage of classical transition state theory (TST) ideas. In classical harmonic TST (Section 8.2.5), the transition rate is determined by the behavior of the system in the vicinity of saddle points of the underlying potential. To extend this idea to quantum systems, we search for the saddle points of the action, Eq. (12.63). Those are stationary points satisfying the equation $\partial S_E / \partial x_n = 0$.

At this point, it is helpful to switch back to the continuous case. "Stationary points" of the Euclidian action $S_E[x(\tau)]$ are the trajectories that correspond to the extrema of the action. Because $S_E[x(\tau)]$ can be viewed as a Lagrangian action for the motion in the upside-down potential (compare Eqs. (12.54) and (12.62)), its extrema must correspond to classical trajectories in the potential $-U(x)$. Those satisfy Newton's second law,

$$m\ddot{x} = U'(x) \qquad (12.65)$$

as well as the cyclic boundary condition $x(0) = x(\beta\hbar)$. We are, therefore, led to consider periodic orbits in the upside-down potential shown in Figure 12.3. There are two trivial solutions to Eq. (12.65) corresponding to the system residing at a maximum $(x(\tau) = x_A)$ and the minimum $(x(\tau) = x^{\ddagger})$ of the upside-down potential. In addition, when the period $\beta\hbar$ exceeds the frequency $\omega^{\ddagger}/(2\pi)$ of small vibrations around $x^{\ddagger}$, a periodic orbit traversing the barrier region back and forth also satisfies both the boundary condition and Equation (12.65). This orbit is known as the "bounce" (or "instanton") solution. This, of course, is the same trajectory as the one we have discussed above (cf. Eq. (12.28)).

It can be shown that the first of the two trivial solutions is always the minimum of the Euclidian action, while the second one is a saddle. The bounce solution is also a saddle. A classical harmonic transition theory (see Section 8.2.5) in the extended space of ring polymers (i.e. discretized versions of periodic orbits $x(\tau)$) is a promising candidate of a quantum transition state theory [7, 9–12]. Above the crossover temperature, the trivial solution leads to a Euclidian action equal to $S_E = \beta\hbar U^{\ddagger}$, leading to the usual Arrhenius law. Quantum corrections, however, are incorporated by this theory once the prefactor is computed, as in harmonic TST. Below the crossover temperature the delocalized bounce solution is the correct transition-state saddle, giving the dominant contribution to the quantum rate. As the temperature is lowered, the bounce amplitude increases, indicating increasing quantum delocalization within the barrier. The same results as above (e.g. Eqs. (12.30) and (12.41)) can be obtained from this picture. The advantage of the path integral formulation is, however, two-fold. First, it allows straightforward generalization to cases of higher dimensionality, where periodic orbits on the inverted PES are considered. Second, it allows powerful numerical tools developed for classical TST to be applied to the quantum case. Those include, for example, algorithms for finding saddle points on multidimensional potential energy surfaces [13].

Other ideas from classical rate theory can be applied within the path integral formulation. For example, one may seek a reduced description of the problem in terms of a low-dimensional reaction coordinate. For a polymer ring an intuitively appealing candidate for such a coordinate is its centroid (Figure 12.5),

$$x_c = N^{-1} \sum_{n=0}^{N} x_n \approx (\beta\hbar)^{-1} \int_0^{\beta\hbar} x(\tau)d\tau. \tag{12.66}$$

The "potential of mean force", $F(x_c)$, as a function of the centroid coordinate, constitutes an effective quantum potential that accounts for quantum delocalization effects [14]. In the centroid theory approach, then, one considers transitions in this quantum potential, instead of the original potential $U(x)$. At high temperature a periodic orbit shrinks to a point, and the quantum potential becomes equal to $U(x)$.

References

1 Landau, L.D. and Lifshitz, E.M. (1980). *Statistical Physics*, vol. 5. Butterworth-Heinemann.

2 Landau, L.D. and Lifshits, E.M. (1965). *Quantum Mechanics: Non-Relativistic Theory*, 2e. Oxford, New York: Pergamon Press.

3 Benderskii, V.A., Makarov, D.E., and Wight, C.A. (1994). *Chemical Dynamics at Low Temperatures*. New York: Wiley.

4 Chandler, D. and Wolynes, P.G. (1981). Exploiting the isomorphism between quantum theory and classical statistical mechanics of polyatomic fluids. *J. Chem. Phys.* 74: 4078–4095.

5 Caldeira, A.O. and Leggett, A.J. (1983). Quantum tunneling in a dissipative system. *Ann. Phys.* 149: 374.

6 Weiss, U. (2018). *Quantum Dissipative Systems*. World Scientific.

7 Richardson, J.O. (2018). Perspective: Ring-polymer instanton theory. *J. Chem. Phys.* 148 (20): 200901.

8 Richardson, J.O. and Althorpe, S.C. (2009). Ring-polymer molecular dynamics rate-theory in the deep-tunneling regime: connection with semiclassical instanton theory. *J. Chem. Phys.* 131 (21): 214106.

9 Mills, G., Schenter, G.K., Makarov, D.E. et al. (1997). Generalized path integral based quantum transition state theory. *Chem. Phys. Lett.* 278: 91.

10 Mills, G., Schenter, G.K., Makarov, D.E. et al. (1998). RAW quantum transition state theory. In: *Classical and Quantum Dynamics in Condensed Phase Simulations*, (eds. B.J. Berne, G. Ciccotti, and D.F. Coker). World Scientific.

11 Makarov, D.E. and Topaler, M. (1995). Quantum transition state theory below the crossover temperature. *Phys. Rev. E* 52: 178.

12 Miller, W.H. (1975). Semiclassical limit of quantum mechanical transition state theory for non-separable systems. *J. Chem. Phys.* 62: 1899–1906.

13 Henkelman, G., Jóhannesson, G., and Jónsson, H. (2002). Methods for finding saddle points and minimum energy paths. In: *Theoretical Methods in Condensed Phase Chemistry*, 269–302. Netherlands: Springer.

14 Voth, G.A., Chandler, D., and Miller, W.H. (1989). *J. Chem. Phys.* 91: 7749.

13

Computer Simulations of Molecular Kinetics: Foundation

13.1 Computer Simulations: Statement of Goals

Atomically detailed molecular simulations provide comprehensive information on the time evolution of complex systems that are described by well-defined dynamics. In the next few chapters, we will consider motions that follow classical mechanics, a topic which is the focus of most of this book. We discuss the inputs to the simulations, the outputs, and how to connect the results of the simulations to experimental observables and to the approximate theories of rates that were discussed in Chapters 8–11. The mechanic behind the simulations is described in Chapter 15.1 and it relies on the theoretical foundation provided in Chapters 1–7.

The basic output of computer simulations is the coordinate vector $\mathbf{x}(t)$. It is a time-dependent vector that follows the motion of all the atoms in the system. Detailed representations in the liquid phase include the explicit coordinates of the solvent molecules. In other representations the solvent is considered implicitly as a dielectric continuum or a stochastic bath. The use of the Langevin equation (Chapter 1) implies the existence of a solvent bath, which is modeled as friction and random force terms. In typical implicit solvation simulation, the model of the dynamic follows equation (1.19), which we solve numerically for $\mathbf{x}'(t)$, and we reproduce here in a vector matrix form for a large number of degrees of freedom

$$\mathbf{M}\frac{d^2\mathbf{x}'(t)}{dt^2} + \gamma\frac{d\mathbf{x}'(t)}{dt} = \mathbf{F}(t) + \xi(t) \tag{13.1}$$

The mass matrix $\mathbf{M}$ is diagonal in the Cartesian representation, γ is another matrix with the friction coefficients, $\mathbf{F}$ is the vector of forces on each degree of freedom and $\xi(t)$ is the vector of random forces at time t. We use $\mathbf{x}'(t)$ to differentiate this set, which does not include solvent coordinates, from $\mathbf{x}(t)$ in which all the coordinates of the atoms in the systems, solvent or solute, are included.

Molecular Kinetics in Condensed Phases: Theory, Simulation, and Analysis,
First Edition. Ron Elber, Dmitrii E. Makarov and Henri Orland.
© 2020 John Wiley & Sons Ltd. Published 2020 by John Wiley & Sons Ltd.

In a more detailed representation, the solvent molecules are explicitly accounted for and are followed as a function of time. Therefore, it is not necessary to include noise and friction as a model for an external bath. The bath is described explicitly. The equations of motion follow Newton's laws. That is, we have

$$\mathbf{M}\frac{d^2\mathbf{x}}{dt^2} = \mathbf{F} \tag{13.2}$$

The goal of atomically detailed simulations is to provide a picture as close as possible to physical reality. Understanding the mechanism is, of course, important but it is not the only goal. Conducting detailed simulations following explicit dynamics can help test approximate rate theories that are easier to analyze, check for a wide range of parameters, and understand concrete molecular mechanisms. Moreover, detailed simulations may replace experiments when laboratory measurements are expensive or difficult to do and the computational model is sufficiently accurate.

From the perspective of a substitute to experiment, Eq. (13.2) is attractive since the use of phenomenological modeling (like the use of friction and random force, see Chapter 1) is avoided. The comparison to experiment is more straightforward. The vector $\mathbf{x}(t)$ can be used "as is" to analyze macroscopic equilibrium and kinetics. The more detailed picture is obtained at the significant cost of computational resources and complexity of analysis. The length of the vector $\mathbf{x}(t)$ is about one hundred to one thousand times longer than the vector $\mathbf{x}'(t)$. The integration in time of the first vector requires significantly more computational resources compared to the second reduced representation.

Setting the computational complexities aside, given the vector $\mathbf{x}(t)$ it is possible to compute experimental observables of the system. Let $A(\mathbf{x}(t), t)$ be an observable of interest. If there is no external force the observables do not depend on time directly and we therefore write $A(\mathbf{x}(t))$, hence the time dependence is through the coordinates. An equilibrium average $\langle A \rangle_{equ}$ is typically computed in a simulation from a series of small displacements, or more specifically as a time average. This averaging relies on the ergodic hypothesis [1] that equates a time average with a phase space average. Phase space averages are computed over all possible values of the coordinates and the momenta as written on the r.h.s. of Eq. (13.3).

$$\langle A \rangle_{equ} = \lim_{t \to \infty} \frac{1}{t} \int_0^t A(t')dt' = \int p(\mathbf{x}, \mathbf{p})A(\mathbf{x}, \mathbf{p})d\mathbf{x}d\mathbf{p} \tag{13.3}$$

where $p(\mathbf{x}, \mathbf{p})$ is the probability density of the phase space point $(\mathbf{x}, \mathbf{p})$. The trajectories offer more than an equilibrium view. Time dependent properties can be estimated (for example) with correlation functions, $C(t)$. Let $A(x(0))$ be the value of the observable at time zero, and $A(x(t))$ the value at a later time t. We measure the correlation between the values of the observables by averaging the expression $-A(x(0))A(x(t))$. If the system is in equilibrium, then the origin of time is irrelevant, and we can just as well write $A(x(t'))A(x(t + t'))$. Using the ergodic hypothesis, we can time-average over the origin of time t' to have

$$\langle A(x(t'))A(x(t + t'))\rangle = \lim_{t'' \to \infty} \frac{1}{t''} \int_0^{t''} A(x(t'))A(x(t + t'))dt' \qquad (13.4)$$

The last equation shows the time dependence of an averaged observable at equilibrium. This may sound strange, since equilibrium is defined as the state in which macroscopic changes of observables in time are zero. However, fluctuations in time occur even if the system is in equilibrium. Moreover, the system may be in a local equilibrium (see Chapter 13.5) and therefore time-dependent transitions between metastable states are possible.

13.2 The Empirical Energy

The force vector, $\mathbf{F}$, in equation (13.2), is minus the gradient of a potential, U, ($\mathbf{F} = -\nabla U$). In a large number of condensed phase simulations, an empirical formula for the potential is used which is given below. The energy landscape of the nuclei is a Born Oppenheimer surface in which the electrons induce an effective potential for the much heavier nuclei [2]. It depends only on the coordinates of the nuclei. The energy is a linear combination of bonded and non-bonded parts [3, 4]. The bonded energy is a sum of bond, (E_b), angle, (E_θ) and torsion (E_ϕ) terms. The non-bonded energy includes electrostatic interactions (E_{elec}), dispersion energies and excluded volume terms (E_{LJ}).

$$U = U_{bonded} + U_{non-bonded}$$

$$U_{bonded} = \sum_b E_b + \sum_\theta E_\theta + \sum_\phi E_\phi$$

$$E_b = k_b(b - b_0)^2 \quad E_\theta = k_\theta(\theta - \theta_0)^2 \quad E_\phi = \sum_n a_n \cos(n\phi_n + \delta_n)$$

$$U_{non-bonded} = E_{elec} + E_{LJ}$$

$$E_{elec} = \sum_{i>j} \kappa \frac{q_i q_j}{\varepsilon r_{ij}} \quad E_{LJ} = \sum_{i>j} \frac{A_{ij}}{r_{ij}^{12}} - \frac{B_{ij}}{r_{ij}^6} \qquad (13.5)$$

The bonded energy summations are conducted over all bonds, angles, and torsions in the system, while the non-bonded summations are (not surprisingly) for all atoms that are not bonded. The b and θ are the bond lengths and angles of the current molecular geometry. The parameters b_0 and θ_0 are the "ideal" bond and angle values for each individual bonded term. The coefficient k_b and k_θ are empirical force constants that control the stiffness of the bond and angle terms. The above parameters are determined from experiments such as vibrational spectroscopy or crystallography. The torsion energy is written as a Fourier series with a_n for an amplitude, n, a bond rotation period, typically two or three, and δ_n, a phase. The non-bonded interactions include coulomb energies between pairs of particles with charges q_i and q_j respectively and a distance r_{ij}. The coefficient κ is a unit bearing parameter and ε is the dielectric constant. The Lennard Jones potential, E_{LJ}, is parameterized by two coefficients for each pair of particles, A_{ij} and B_{ij}.

Many additions and variations to the basic functional form of the potential are possible. For example, the inclusion of electrostatic polarization [5]. However, a discussion about these enhancements is beyond the scope of the present book.

At room temperature, thermal fluctuations cause deviations of bond lengths from ideal values by about one percent (~ 0.03Å). The fluctuations of the angles from their average positions are less than 10 percent (~ 10 degrees). The stiff potentials prevent larger fluctuations at room temperature. The small thermal displacements of bonds and angles are rapidly changing as a function of time (with a period shorter than 1 picosecond).

Significant variations in molecular geometry can be observed upon changes in torsion angles or rotation around bonds (curve lines in Figure 13.1). The rotation around a single bond (a curve line) is frequently observed since a typical barrier for a single bond rotation is less than the thermal energy at room

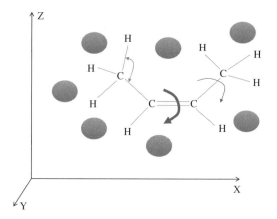

Figure 13.1 A schematic drawing of a butene molecule in solution. The dark spheres denote solvent molecules and the butene molecule is drawn explicitly. Single and double lines denote single and double chemical bonds respectively. A bond angle for the angle CCH is illustrated by a curve with double arrows. Also shown are two torsions that represent rotations around bonds. The thin line is a rotation around a single bond and the thick line a rotation around a double bond. The molecular geometry can be determined by a set of bonds, angles, and torsions, or with the Cartesian positions (X, Y, Z coordinates) of all the atoms in the system. In computer simulations, the Cartesian representation is general, convenient, and most widely used.

temperature, RT. R is the gas constant and T is the temperature in Kelvin degrees. At room temperature ($\sim300°K$) we have $RT = 300 \times 0.002 = 0.6 kcal/mol$. In contrast, the barrier for double bond rotation (a thick line) is at least ten times larger, making it significantly less common to observe. Infrequent transitions of this type are called in the field "rare events" and are targets of numerous simulation technologies.

13.3 Molecular States

In this section, we define molecular states that guide simulation technologies (see also Chapter 8.1 on isomerization process). We use molecular states to compute the progress of molecular events at the appropriate timescales. We start with a simple example, which is in the same spirit of the chair-boat isomerization that is discussed in Chapter 8.1. The text below adds, however, the concept of local roughness and separation of time scales. Consider a 2-butene molecule, $CH_3CH = CHCH_3$ which we call butene from now on. The butene molecule has one double bond between the second and the third carbon and single bonds between other atoms as sketched in Figure 13.1.

The double bond rotation switches the molecule from a cis conformation (the two CH_3 groups are on the same side of the double bond) to a trans configuration (the two CH_3 groups are at opposite sides of the double bond). The process is similar in spirit to the chair boat transition depicted in Chapter 8.1 of a two-state system. The single bond rotation (rotations of the hydrogen positions of the CH_3 groups) impacts slightly the energy of the double bond rotation. The energy landscape can be modulated further by non-bonded interactions with the solvent molecules (blue spheres in Figure 13.1). For example, a hard-core collision of a rotating atom i with a solvent molecule j (the A_{ij}/r_{ij}^{12} term) will increase the barrier for rotation while long-range attraction (the $-B_{ij}/r_{ij}^6$ term) reduces the overall energy and may reduce the barrier as well.

The discussion above suggests a distribution and a hierarchy of barrier heights in condensed phase simulation. A one-dimensional energy profile for the rotation of the double bond of solvated butene is sketched in Figure 13.2.

We call the collection of all the atoms (e.g. butene and solvent molecules) the system. Consider an experiment that probes the spatial and temporal behavior of the system. It can be a computer experiment or an experiment in a "wet" laboratory. At maximum resolution the coordinates of all the atoms in the system are recorded in the vector $\mathbf{x}(t)$, for each instance of time, t. The vector $\mathbf{x}(t)$ is called a *trajectory* and it describes the *dynamics* of the system. In a typical "wet" laboratory experiment, however, the resolution is less than maximal. Let us assume that only the torsion value of the double bond, ϕ, is recorded as a function of time. Figure 13.3 sketches a plausible time series of ϕ.

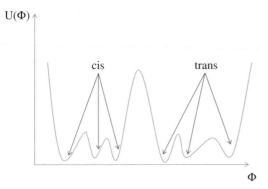

U(Φ)

cis   trans

Φ

Figure 13.2 A sketch of the energy profile $U(\phi)$ of a solvated butene as a function of the torsion angle ϕ. The small energy bumps correspond to rotations of single bonds. These energies are also impacted by packing of solvent molecules against the butene and along Φ. At thermal energies we expect rapid transitions over the small energy bumps and significantly slower transitions between the cis and trans configurations.

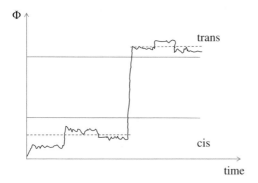

Φ

trans

cis

time

Figure 13.3 A sketch of the changes in the torsion angle of the double bond in butene as a function of time. The middle horizontal lines separate the cis and the trans configurations. The trans is the upper, and the cis is the lower part of the graph. The system spends most of its time either at the cis or the trans states. If the resolution is low, we may not observe the transitions over the small energy bumps of Figure 13.2 or identify Φ values during the transition between cis and trans. In the low-resolution case, we only differentiate between the binary cis and trans states as a function of time (dashed line). Figure 13.3 depicts the same type of an activated process and separation of time scales and barrier heights that are discussed in figure VIII.2.

If the recording is of a lower resolution (in time and/or space) our data may look like the dashed line of Figure 13.3 in which only averaged Φ values within the cis and trans states are observed. We comment that a limited spatial and temporal resolution can be useful. It suggests a simpler description of the time evolution of a single butene molecule using only the most relevant degree of freedom. Sometimes all the details are not desired.

In the laboratory, we rarely measure the time series of a single molecule (though single molecule experiments are becoming more wide spread) [6, 7]. We usually measure the time courses of a large number of molecules. For example, instead of reporting the cis or trans states of individual molecules, the experiment returns the number of molecules that are in the cis state at time t.

We call the reduced representation in time and space of the molecular system averaged over an ensemble of dynamic events *kinetics*.

13.4 Mean First Passage Time

The MFPT (Mean First Passage Time) or $< \tau >$ is the averaged transition time between two states using a sample of statistically independent trajectories. We are using here the perspective of individual trajectories. See also Chapter 6.2.2 for a discussion about the MFPT in the context of Kramer's equation. It is measured from initiation at one state until the trajectory enters for the first time into the other state. In our concrete example of the butene molecule the initiation state can be the cis state and the second state to which the trajectory enters for the first time is the trans. The MFPT is a prime observable for kinetics and is reported, directly or indirectly, in numerous experiments and theoretical studies. It captures essential characteristics of the kinetics and therefore has received considerable attention. We define the MFPT in Eq. 13.6, below and discuss it in numerous places in this book.

$$\langle \tau \rangle = \frac{1}{L} \sum_{i=1}^{L} t_i \tag{13.6}$$

L is the number of reactive trajectories and t_i is the first passage time of trajectory i.

The first passage time is a stochastic variable with a probability density $\pi(\tau)$. The MFPT is the first moment of this distribution. See also Equation VIII.7 for the dwelling time. It provides a useful characterization of the time of the reaction. However, this description is clearly not complete. To fully characterize the kinetics, the distribution, $\pi(\tau)$, or all the moments of it, such as the second moment - $\langle \tau^2 \rangle = \int \pi(\tau)\tau^2 d\tau$ are required. In typical biochemical and biophysical applications, the focus is on the rate coefficient, k, which is expressed as $1/\langle \tau \rangle$ under some conditions (see Chapter 14).

13.5 Coarse Variables

In experimental studies of kinetics, we rarely measure the system dynamics at maximum spatial resolution. We compromise instead on probing the values of a small number of coordinates or degrees of freedom, which (we hope) gives adequate description of the time evolution of the whole system. The subsets of coordinates, which we use to probe the kinetics, are called *coarse variables*. In the case of butene we consider the torsion angle ϕ as a coarse variable. Our discussion is, of course, not limited to butene or to torsion angles. The coarse variables can differ widely depending on the experimental techniques and the system of interest. Typically, they are a single or a combination of the internal coordinates, which we introduce in 13.2: bonds, angles, or torsions. However, they can also be more global variables. An example is the radius of gyration as measured by SAXS experiments [8]. The radius of gyration is a convenient probe to follow the collapsed state of a polymer or a protein. The coarse variables may also include the number of protected hydrogen bonds as detected in hydrogen exchange experiments during protein folding [9]. They can also be more macroscopic entities like volume or the surface area of a macromolecule. We may use one or several coarse variables to define a *reaction space*. We call a subset of variables denoted by $\mathbf{y}$ a *reaction space* if these variables are needed to follow the progress of the reaction. In general, there is one to one transformation from the full coordinate set $\mathbf{x}$ to $\mathbf{y}$ but the reverse is not true.

13.6 Equilibrium, Stable, and Metastable States

Equilibrium is defined as a state of the system in which averaged observables are time independent. It is necessary to use an average over an ensemble of molecules to define equilibrium since a trajectory $\mathbf{x}(t)$ of a single molecule never stops moving at finite temperatures (e.g. Figure 13.3). Only the averages may become time independent. Drawing again on the butene example, we consider an equilibrium state of the cis and the trans conformations. A fraction of the butene molecules are in the cis state and some of these molecules will transition in the next time interval to the trans state. Hence, microscopically, when examining one molecule at a time, we observe that the molecular state is time dependent. However, when the system is in equilibrium, molecules that are on the move from trans to cis are balanced by the molecules that transition from cis to trans. The net change in the populations of both states is zero and the state of the system averaged over the population is time independent. Statistical mechanics considers a system to be in equilibrium if the *probability* of being in any state is time independent.

In classical mechanics, a phase-space vector of coordinates and momenta of all the atoms $(\mathbf{x}, \mathbf{p})$, determines a "pure" state of the system, which we also call a precise state (PS) since it is a state of maximum resolution. For a complete and general description of the system, in equilibrium or not, we need the probability densities of all the PS-s as a function of time, $P(\mathbf{x}, \mathbf{p}, t)$. At equilibrium, we only need a time-independent probability $P_{eq}(\mathbf{x}, \mathbf{p})$. In the canonical ensemble (configurations sampled at a constant volume, number of particles, and temperature) we know that the probability density of a PS is

$$P_{eq,canonical}(\mathbf{x}, \mathbf{p}) = \frac{\exp(-\beta H(\mathbf{x}, \mathbf{p}))}{Z'}$$

$$Z' = \int d\mathbf{x}d\mathbf{p} \cdot \exp(-\beta H(\mathbf{x}, \mathbf{p}))$$

$$H(\mathbf{x}, \mathbf{p}) = \frac{1}{2}\mathbf{p}^t\mathbf{M}^{-1}\mathbf{p} + U(\mathbf{x}) \tag{13.7}$$

where $\beta = 1/k_B T$ and k_B is the Boltzmann constant. We write the Hamiltonian $H(\mathbf{x}, \mathbf{p})$ with a coordinate-independent mass matrix, $\mathbf{M}$. This is the case when Cartesian coordinates and momenta are used. We reserved the symbol Z for the configuration integral $Z = \int d\mathbf{x} \cdot \exp(-\beta U(\mathbf{x}))$ and therefore use Z' for normalization when the weight is determined by the full Hamiltonian including momentum average.

Consider a state α, which is a part of the whole phase space accessible to the system. An example is the cis configuration of the solvated butene that includes many precise states that are still cis. That is, we retain the cis configuration of the double bond but consider the different configurations of the solvent molecules, rotational states of the CH_3 groups, etc., as part of the α state. The state α is said to be metastable if the relative probabilities of the PS-s within the state are those of equilibrium. The state is not necessarily stable since the overall population of state α may be time dependent. Physically speaking a metastable state is observed when there is a separation of time scales. Barrier crossing and relaxation to equilibrium are much faster within the metastable state compared to transitions between different metastable states. Alternatively, we say that the metastable state α is in Local Equilibrium (LE). More formally, we write

$$P_\alpha(\mathbf{x}_\alpha, \mathbf{p}_\alpha, t) = P_{\alpha,LE}(t)P_{\alpha,equilibrium}(\mathbf{x}_\alpha, \mathbf{p}_\alpha) = P_{\alpha,LE}(t)\frac{\exp(-\beta H(\mathbf{x}_\alpha, \mathbf{p}_\alpha))}{Z'_\alpha}$$

$$Z'_\alpha = \int d\mathbf{x}_\alpha d\mathbf{p}_\alpha \exp(-\beta H(\mathbf{x}_\alpha, \mathbf{p}_\alpha)) \tag{13.8}$$

Where the integration is restricted to the phase space points that belong to state α. The set of metastable states is assumed complete, that is, $Z' = \sum_\alpha Z'_\alpha$.

Identifying metastable states is a topic of considerable research. Classification of metastable states was investigated within the framework of Macrostates [10], and Markov State Models [11, 12]. Shalloway [10] proposed time independent analysis of the peaks in the equilibrium distribution to determine the metastable collections of PS-s. In the Markov State Model, we identify states that are separated by time scale gaps [11, 12]. We call these states macrostates, or metastable states, and in short MS-s.

In summary, the local equilibrium assumption is valid in the limit in which the transition time to other MS-s is much slower than the relaxation times within the MS leaving ample time for the system to relax into a local equilibrium. This difference in barrier heights and therefore in time scales is sketched in Figure 13.2. This separation is highly convenient from computational and analysis points of views. However, it is not always possible. For example, when the energy landscape is relatively flat without high barriers to differentiate between the MS-s, no simple definition of MS-s is possible. In other words, there must be a significant gap between the time scale required to cross the highest barriers and the time scale for crossing the next set of barriers.

References

1 McQuarrie, D. (2000). *Statistical Mechanics*. Sausalito: University Science Books.

2 Cohen-Tannoudju, C., Diu, B., and Laloe, F. (1977). *Quantum Mechanics*. New York: Wiley.

3 McCammon, J.A. and Harvey, S.C. (1987). *Dynamics of Proteins and Nucleic Acids*. Cambridge: Cambridge University Press.

4 Schlick, T. (2002). *Molecular Modeling and Simulation: An interdisciplinary Guide*. New York: Springer Verlag.

5 Ren, P.Y. and Ponder, J.W. (2003). Polarizable atomic multipole water model for molecular mechanics simulation. *J. Phys. Chem. B* 107 (24): 5933–5947.

6 Gell, C., Brockwell, D., and Smith, A. (2006). *Handbook of Single Molecule Spectroscopy*. Oxford: Oxford University Press.

7 Makarov, D.E. (2015). *Single Molecule Science: Physical Principles and Models*. CRC Press, Taylor & Francis Group.

8 Pollack, L. (2011). SAXS studies of ion-nucleic acid interactions. In: *Annu. Rev. Biophys.* (eds. D.C. Rees, K.A. Dill, J.R. Williamson) 40: 225–242.

9 Englander, S.W., Mayne, L., Kan, Z.Y. et al. (2016). Protein folding-how and why: by hydrogen exchange, fragment separation, and mass spectrometry. In: *Annu. Rev. Biophys.* (ed. K.A. Dill) 45: 135–152.

10 Shalloway, D. (1996). Macrostates of classical stochastic systems. *J. Chem. Phys.* 105 (22): 9986–10007.

11 Schutte, C. and Sarich, M. (2013). *Metastability and Markov State Models in Molecular Dynamics: Modeling, Analysis, Algorithmic Approaches*, vol. 24. American Mathematical Society.

12 Bowman, G.R. and Pande, V.S. (2014). *An Introduction to Markov State Models and Their Applications to Long Timescale Molecular Simulations*. Springer, p. 139.

14

The Master Equation as a Model for Transitions Between Macrostates

In this section we consider a widely used model to investigate transitions between multiple MS-s, the Master Equation. Formal properties of this model are discussed in this book in Chapter 4. For additional discussion on first order chemical kinetics see Chapter 8.1. Here, we provide a somewhat different picture of the Master Equation that by the end of the day may lead to more efficient computer simulations.

The rate of transition from MS α to MS β is the fraction of the population that moves from state α to state β in unit time. A phenomenological expression for the rate is $rate(\alpha \rightarrow \beta) = k_{\alpha \rightarrow \beta} P_{\alpha, LE}(t)$. For those familiar with chemical kinetics, [1] do not confuse this expression with the equations of chemical kinetics, in which we write (for example) $rate = k[A][B]$ where [A] and [B] are concentrations of chemical species. The metastable and abstract state α is a complex characterization of the whole system in phase space. The formulae of chemical kinetics are a drastic simplification assuming, for example, a uniform distribution of molecules A and B in the reaction volume. State α can be a uniform mixture of molecules A and B, but it can also include spatial correlations in the volume of the reaction and other reaction descriptors.

The rate coefficient $k_{\alpha \rightarrow \beta}$ is independent of the state probability, is non-negative, and is derived using an equilibrium distribution within the MS α. It is obviously critical for our understanding of kinetics. In this section, we do not compute it and consider it given. In Chapter 15 we discuss how the rate coefficients between individual MS-s are calculated from simulations of PS dynamics. Approximate and exact theories for estimating the rate coefficients are found in Chapter 8. The phenomenological rates are combined to provide the rate of change of the population of a state α, $dP_{\alpha, LE}(t)/dt$

$$\frac{dP_{\alpha,LE}(t)}{dt} = \sum_{\beta} -k_{\alpha \rightarrow \beta} P_{\alpha,LE}(t) + k_{\beta \rightarrow \alpha} P_{\beta,LE}(t) \tag{14.1}$$

Eq. (14.1) is equivalent to Equation (IV.9) written here for metastable states and is called the Master Equation (ME). The first term on the r.h.s. describes

Molecular Kinetics in Condensed Phases: Theory, Simulation, and Analysis,
First Edition. Ron Elber, Dmitrii E. Makarov and Henri Orland.
© 2020 John Wiley & Sons Ltd. Published 2020 by John Wiley & Sons Ltd.

loss of population from state α, the second term a gain. A process that follows the ME is Markovian. A modern approach for the analysis of simulations is called the Markov State Model [2] and it employs extensively formulations of the type of Eq. (14.1).

A more common definition of a Markovian process is using a conditional probability. Suppose we are given an ordered sequence of configurations generated by a prespecified process (trajectory) $\{x_1, x_2, \ldots, x_n\}$. We ask that "given the sequence, what is the probability of observing x_{n+1} as the next configuration?". We can write in general the (conditional) probability as $p(x_{n+1} \mid x_1 \ldots x_n)$. A process is Markovian if the condition extends only to the previous configuration: $p(x_{n+1} \mid x_n) = p(x_{n+1} \mid x_1 \ldots x_n)$. In general, classical trajectories of PS described, for example, by Eq. (13.2) are Markovian. However, metastable states that are modelled by trajectory configurations that are clustered to MS are not necessarily Markovian and their kinetics are not necessarily captured by an ME.

Suppose that all the population is concentrated at time zero at state α, hence $P_{\alpha, LE}(0) = 1$. After a very short time there is a flow of probability from state α to other states. Since the other states are still empty there is no flow back into state α from other MS states. Alternatively, we can set all the states β that are directly connected to α to be absorbing and terminating for the purpose of this conceptual experiment. By "directly connected" or "directly accessible" we mean that a trajectory can be found that transitions from state α to state β without visiting another state along the way. By "absorbing and terminating" we mean that every trajectory that leaves state α disappears when it enters state β. Therefore, there are no trajectories that return to α after entering β and the population of state α must be monotonically decreasing. If the ME is followed, we have under these conditions.

$$\lim_{t \to 0} \frac{dP_{\alpha, LE}(t)}{dt} = -\sum_{\beta} k_{\alpha \to \beta} P_{\alpha, LE}(t) \qquad (P_{\alpha, LE}(0) = 1)$$

$$\lim_{t \to 0} P_{\alpha, LE}(t) = \exp\left[-\sum_{\beta} k_{\alpha \to \beta} t\right] = \exp(-\Lambda t) \qquad (14.2)$$

The final expression shows a single exponential relaxation in time with an effective rate coefficient, Λ, which is the sum of rate coefficients to all other states, $\Lambda = \sum_{\beta} k_{\alpha \to \beta}$. This single exponential decay in time of the MS population to nearby states is a fingerprint of a Markovian process and can be used as a test if the ME models accurately the system under consideration. We may observe the decay of a population of a state by different means, experimentally and by simulations. In both cases, the relaxation rate can be extracted empirically with a straightforward analysis. For example, plotting $\log[P_{\alpha, LE}(t)]$ as a function of t gives a straight line with a slope of $-\sum_{\beta} k_{\alpha \to \beta}$ at the short time limit (or satisfying the absorbing and terminating conditions).

A frequent question in kinetics is "Can we model the dynamics using a **Markovian** approach (or a Master equation) with a smaller number of degrees of freedom and avoid the full phase space of the system"? If the populations of each of the states, which we define or measure, decay exponentially in the short time limit as depicted in Eq. (14.2) then the states are Markovian and the answer to the question at the beginning of the paragraph is "yes".

We can write Eq. (14.1) in a more compact form. We consider the matrix **k** with off-diagonal elements the rate coefficients, $k_{\alpha \to \beta}$ and diagonal elements as minus the sum of these coefficients, $-\sum_\beta k_{\alpha \to \beta}$. The indices of the matrix are, of course, the MS-s.

$$\frac{d\mathbf{P}}{dt} = \mathbf{k}\mathbf{P} \tag{14.3}$$

At equilibrium state(s), which are time independent and are less interesting from kinetic perspective, the time derivatives of the populations of the metastable states are zero. The time-independent states are solution of the linear equations for the equilibrium probabilities, $P_{\alpha, eq}$, of each of the metastable states α, $\left[\sum_\beta (\mathbf{kP})_\beta\right]_\alpha = -\sum_\beta k_{\alpha\beta} P_{\alpha,eq} + \sum_\beta k_{\beta\alpha} P_{\beta,eq} = 0$. Hence at least one eigenvalue of **k** (the equilibrium state) must be zero with a corresponding non-zero eigenvector.

The ME offers a simple solvable model for a number of functions of wide interest, which were discussed in Chapter 4. We will illustrate some applications using a useful tool, the transition probability $\mathbf{w}(t)$, which we also call the kernel. It is a matrix with elements $w_{\alpha\beta}(t)$. Each element is the probability that the system transitions directly to state β between t and $t+dt$ given that the system is in a state α at time t. We note that the transition probability or the kernel can be used broadly. It can be exploited to investigate many types of dynamics. It is not limited to Markovian processes, even though we illustrate it below for the ME. The kernel is normalized as follows:

$$\sum_\beta \int_0^\infty w_{\alpha\beta}(t)dt = 1 \quad w_{\alpha\alpha} = 0 \tag{14.4}$$

We implicitly assume that the state α can transition to at least one other state. If it cannot, the system is trapped in state α and there is no kinetics to consider. For an example, it is useful to consider an explicit realization of the kernel using the Master equation (Eq. (14.1)). $w_{\alpha\beta}(t)dt$ is determined by a loss from state α to state β at time t:$-dP_{\alpha \to \beta}(t)$ under the condition of an absorbing β state (i.e. no back transitions from the β states to α). From Eq. (14.2) we write$-dP_{\alpha\to\beta}(t) = k_{\alpha\beta}P_\alpha(0)\exp\left[-\sum_\gamma k_{\alpha\gamma} t\right] dt$

Setting $P_\alpha(0) = 1$ we obtain an expression for the kernel of the ME

$$w_{\alpha\beta}(t) = k_{\alpha\beta}\exp\left[-\sum_\gamma k_{\alpha\gamma} t\right] \tag{14.5}$$

Note that the kernel defined in equation (14.5) is normalized properly (check Eq. (14.4)). We write

$$w_{\alpha\beta} = \int_0^\infty w_{\alpha\beta}(t)dt = \frac{k_{\alpha\beta}}{\sum\limits_{\gamma} k_{\alpha\gamma}} \tag{14.6}$$

which is the transition probability from α to β summed up over all times given that all the β states are absorbing. Eq. (14.6) is particularly useful to answer how the population initially placed at α splits between the directly accessible states.

What can we do with the transition probability? An interesting example is the calculation of the committor $\mathbf{C}_\alpha$. The committor is the probability that a trajectory initiated at state α will enter a pre-determined product state before entering a reactant state. I.e., it is the level of commitment of a state α to react. Note that the question focuses on yield and not on rate. The committor is time independent. The set of states with the same value of the committor are called iso-committor surfaces. They define an "optimal" reaction coordinate [3] as a sequence of surfaces with a monotonically increasing committor value from zero at the reactant to one at the product. The transition probability we just discussed can be used to compute this optimal reaction coordinate. The discussion bellow follows reference. [4]

Consider the following adjustment of the transition probability. We set the states of both the reactant and the product to be absorbing. Hence, every trajectory that enters the reactant or the product does not exit from these states. The reactant state is also set to be terminating. Every trajectory that enters the reactant disappears. In contrast, the product retains every trajectory that "touches" its boundary. It just does not let it go anymore.

To illustrate these concepts and the calculation of the committor we examine a simple model of only four states (Figure 14.1)

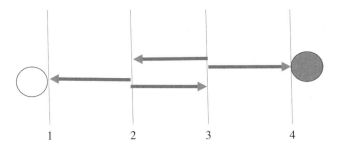

1   2   3   4

Figure 14.1 An illustration of transitions between four discrete states. State 1 is of the reactant and state 4 is the product. The reactant and the product are absorbing while the reactant is also terminating. Every trajectory that enters state 1 disappears, while a trajectory that makes it to the product is getting "stuck" in that state.

A concrete example of a kernel for the scheme of Fig 14.1 is the matrix of Eq. (14.7).

$$w = \begin{pmatrix} 0 & 0 & 0 & 0 \\ p & 0 & 1-p & 0 \\ 0 & p & 0 & 1-p \\ 0 & 0 & 0 & 1 \end{pmatrix} \tag{14.7}$$

Where p, the probability of transition, is a number between zero and one. For example, state 2 transitions to state 1 with probability p and to state 3 with probability $(1-p)$. Note that the matrix **w** must be fully connected with the exception of the first and last columns (reactant and product states) to enable transitions between all the states. If we multiply a row probability vector $\mathbf{p}^t$ by the matrix **w** from the right we obtain the probabilities of the states after one jump to the nearest neighbors, or the splitting of the probability initially at α between nearby states. A single multiplication is equivalent to a transition at any time from the current state to nearby neighbors that are directly accessible to it and are made absorbing.

We keep multiplying the resulting vector by the transition matrix. Each time we jump to the nearby states. After a potentially large number of vector-matrix multiplications, the vector converged to a stationary value in which the remaining trajectories are all trapped in the product state. If the initial probability vector $\mathbf{p}^t$ is not a pure reactant or product, and the reactant and product states are both accessible via multiple multiplications from any other state, a fraction of the probability will get stuck at the product state and the remaining fraction will disappear or terminate at the reactant state.

The final product fraction is an accumulation of *first hitting time distribution*. A first hitting distribution, FHD_α, is the distribution of all trajectory termination points that "touch" state α for the first time. i.e. no re-crossing to other states and returning to α are counted. Since the product is absorbing, all the trajectories that hit the product state must form an *FHD*. This distribution enhances the accuracy of the Transition State Theory (TST) (see Chapter 8). This is since a backward trajectory initiated from the *FHD*, will not include re-crossing events. Of course, the calculation of the first hitting distribution, without the use of explicit trajectories is not trivial.

It may not be obvious what the time-independent multiplications by the kernel means for physical processes. It is, perhaps, easier to understand it by considering a single trajectory. The equivalence between a Monte Carlo procedure representing one trajectory and the Master Equation was discussed in Chapter 4. If a single trajectory starts at state α, it must transition to one of the nearby states. Suppose it made it to state β with probability $w_{\alpha\beta}$, where will it go next and what is the probability of making that transition? Since it is now at state β the probability that it will move to another state γ is $w_{\beta\gamma}$. Hence the probability of a realization of a single trajectory is a sequence of products of $w_{\beta\gamma}$. The product describes transitions between metastable states that the single trajectory visited before it was terminated - $w_{\alpha\beta} w_{\beta\gamma} \ldots w_{\delta s}$ where s is one of the absorbing states (reactant or product). An ensemble of trajectories is sampled according to the probability p_α of the initial conditions of each of the trajectories and transitions to other states are summed. For example, if the first trajectory hops from α to β, a second trajectory may hop from α to β'. It is necessary to sum over all outcomes.

$$p_{traj}(s) = \sum_\alpha p_\alpha \left[\sum_\beta w_{\alpha\beta} \left(\sum_\gamma w_{\beta\gamma} \cdots \left\{ \sum_\delta w_{\delta s} \right\} \right) \right] \tag{14.8}$$

The probability $p_{traj}(s)$ is of a trajectory that ends up in one of the terminating states s.

We can write the above expression as n sequential vector matrix multiplications

$$\mathbf{p}_{traj}(\mathbf{s}) = \mathbf{p} \mathbf{w}^n \tag{14.9}$$

The expression (14.9) suggests that rather than focusing on the vector it is convenient to focus on the matrix and consider $\lim_{n \to \infty} \mathbf{w}^n$. The 4x4 matrix of the type of Eq. (14.7) is reduced by multiplication to

$$\lim_{n \to \infty} \mathbf{w}^n = \begin{pmatrix} 0 & 0 & 0 & 0 \\ 0 & 0 & 0 & C_2 \\ 0 & 0 & 0 & C_3 \\ 0 & 0 & 0 & 1 \end{pmatrix} \tag{14.10}$$

A vector with the values of the iso-committor for each state is obtained by multiplying Eq. (14.10) from the right by a vector of ones.

$$\lim_{n \to \infty} \mathbf{w}^n \mathbf{1} = \begin{pmatrix} 0 & 0 & 0 & 0 \\ 0 & 0 & 0 & C_2 \\ 0 & 0 & 0 & C_3 \\ 0 & 0 & 0 & 1 \end{pmatrix} \begin{pmatrix} 1 \\ 1 \\ 1 \\ 1 \end{pmatrix} = \begin{pmatrix} 0 \\ C_2 \\ C_3 \\ 1 \end{pmatrix} \tag{14.11}$$

If we start at the reactant we will never make it to the product, and therefore the committor value is zero. If we start at the product we stay there with probability of one. As a numerical example we set $p = 1/2$ for the transition matrix of Eq. (14.6). We obtain $\mathbf{C}^t = (0, 0.33, 0.67, 1)$. Hence a process that starts from state 2 has 0.33 probability of making it to the product instead of the reactant. Computationally, the converged vector was obtained by multiplying the matrix of Eq. (14.7) four times by itself using single precision arithmetic. The number of required multiplications will depend on the difference between the largest eigenvalue of **w** (which is one) and the next largest eigenvalue.

Another quantity for which we can derive a closed form solution using the transition probability is the Mean First Passage Time (MFPT – see Eq. (13.6) for definition). The MFPT can be measured experimentally in contrast to the committor, which makes it particularly interesting. We write a linear equation for the MFPT with the knowledge we have acquired so far. We consider the MFPT from state α to state β, which we denote by $<\tau_{\alpha\beta}>$. Let γ be a state that is accessible directly from α. For example, in our simple model shown in Figure 14.1, state 3 is directly accessible from state 2, but state 3 is not directly accessible from state 1. The MFPT from state γ to β is denoted by $<\tau_{\gamma\beta}>$. Let the time of a single and direct transition event between α and γ be t. The First Passage Time (FPT), $\tau_{\gamma\beta}$, is a random variable that does not depend on t or on $w_{\alpha\gamma}$. The time of a single transition event from α to β is a sum of two independent random variables: $\tau_{\alpha\beta} = t + \tau_{\gamma\beta}$. To obtain $<\tau_{\alpha\beta}>$ we average over all directly connected intermediate states γ and intermediate times t.

$$\langle \tau_{\alpha\beta} \rangle = \sum_{\gamma} \int_0^{\infty} t \cdot w_{\alpha\gamma}(t) dt + \sum_{\gamma} w_{\alpha\gamma} \langle \tau_{\gamma\beta} \rangle$$

$$\langle \tau_{\alpha\beta} \rangle = t_{\alpha} + \sum_{\gamma} w_{\alpha\gamma} \langle \tau_{\gamma\beta} \rangle \tag{14.12}$$

where t_{α} is the lifetime of state α. Note that we average this time over all the direct γ exit channels from state α. Eq. (14.12) has the following verbal interpretation. The MFPT from states α to β is given by a sum of the transition times from state α to directly connected states γ plus the MFPT from the states γ to β. We can write Eq. (14.12) in a vector-matrix form

$$(\mathbf{I} - \mathbf{w})\langle \tau_{\beta} \rangle = \mathbf{t} \quad \langle \tau_{\beta} \rangle = (\mathbf{I} - \mathbf{w})^{-1} \mathbf{t} \tag{14.13}$$

where τ_{β} is the vector of all the MFPT-s from any state to state β. Note also that we set $w_{\beta\beta} = 0$. This is different from the calculations of the committor in which $w_{\beta\beta}$ is set to 1. Currently, the trajectories are terminated once they enter the last product state. This is necessary, since we measure the lifetime of the trajectories until they make it to the product β. If we do not terminate them after arrival to β, they will live forever and the MFPT will diverge.

Eq. (14.12) is very general and can be applied to all processes for which we can define a time dependent transition matrix **w**(t) (derived from the ME, or not). It is, nevertheless, instructive to write an explicit expression for the MFPT using the ME. The lifetime is given by

$$t_{\alpha} = \sum_{\gamma} \int_0^{\infty} t \cdot k_{\alpha\gamma} \exp\left[-\sum_{\gamma} k_{\alpha\gamma} t \right] \cdot dt = \frac{1}{\sum_{\gamma} k_{\alpha\gamma}} \tag{14.14}$$

Equation (14.14) is a simple generalization of Equation (VIII.7), which was written for a two-state system. Eq. (14.12) can be written more explicitly for the Master Eq.

$$\langle \tau_{\alpha\beta} \rangle = \frac{1}{\sum_{\gamma} k_{\alpha\gamma}} \left(1 + \sum_{\gamma} k_{\alpha\gamma} \langle \tau_{\gamma\beta} \rangle \right) \tag{14.15}$$

Eq. (14.15) is remarkably simple. The MFPT is determined from a linear equation with the rate coefficients of the Master Eq. as input parameters.

Eq. (14.1) is a phenomenological equation for MS-s. Only if the states α are in local equilibrium, or are made so small that the MS is the same as PS, i.e. a phase space point $(\mathbf{x}, \mathbf{p})$, the ME can be made exact. Nevertheless, it is possible to adjust Eq. (14.1) to obtain a rigorous formulation for MS-s in general. Instead of the ME we have the Generalized Master Equation (GME [5])

$$\frac{d\mathbf{P}}{dt} = \int_0^t \mathbf{k}(t - t') \mathbf{P}(t') dt' \tag{14.16}$$

The bold face letters denote vectors and matrices with elements that are the states α. Here the rate coefficients depend on time. To recover the usual ME we set $\mathbf{k}(t) = \mathbf{k} \cdot \delta(t)$, where $\mathbf{k}$ is a matrix of the time-independent rate coefficients and $\delta(t)$ is the Dirac's delta function in time. [6] We note that the time dependent rate coefficients correspond to a non-Markovian process in time. The transition probability depends not only on the previous step, but also on the history of the trajectory given the general functional form of $\mathbf{k}(t - t')$. The GME was introduced by Mori [5]. The monograph by Zwanzig [7] provides additional insights into the GME.

The formulae we derived for the transition kernel and the MFPT are also valid for systems that follow Eq. (14.16). The kernel, $w_{\alpha\beta}(t)$, which we discussed earlier, is a useful function for non-Markovian dynamics as well (e.g. GME). The relationship between the kernel and the rate coefficients is, however, more complex for the GME compared to Eq. (14.5). It relies on the Laplace transforms of the transition kernel and the rate coefficient. The Laplace transform of a function $f(t)$ is defined as

$$\hat{f}(\lambda) = \int_0^\infty \exp(-\lambda t) f(t) dt \tag{14.17}$$

Here we only cite the relationship between $\hat{w}_{\alpha\beta}(\lambda)$ and $\hat{k}_{\alpha\beta}(\lambda)$ which was derived in [8]

$$\hat{k}_{\alpha\beta}(\lambda) = \frac{\lambda \hat{w}_{\alpha\beta}(\lambda)}{\left(1 - \sum_{\beta} \hat{w}_{\alpha\beta}(\lambda) \right)} \tag{14.18}$$

As an illustration consider the transition probability and the rate coefficients from the Master Equation. We have

$$\widehat{k}_{\alpha\beta}(\lambda) = \int_0^\infty k_{\alpha\beta}\delta(t)\exp[-\lambda t]dt = k_{\alpha\beta}$$

$$\widehat{w}_{\alpha\beta}(\lambda) = \int_0^\infty k_{\alpha\beta}\exp\left[-\sum_\beta k_{\alpha\beta}t\right]\exp(-\lambda t)dt = \frac{k_{\alpha\beta}}{\lambda + \sum_\beta k_{\alpha\beta}} \quad (14.19)$$

Substituting $\widehat{w}_{\alpha\beta}(\lambda)$ in the expression for $\widehat{k}_{\alpha\beta}(\lambda)$ (Eq. (14.18)) we perform a "sanity" check to verify the identity expressed using Laplace transforms

$$k_{\alpha\beta} = \frac{\lambda\frac{k_{\alpha\beta}}{\lambda+\sum_\beta k_{\alpha\beta}}}{\left(1 - \sum_\beta \frac{k_{\alpha\beta}}{\lambda+\sum_\beta k_{\alpha\beta}}\right)} = \lambda\frac{\frac{k_{\alpha\beta}}{\lambda+\sum_\beta k_{\alpha\beta}}}{\lambda/\left[\lambda+\sum_\beta k_{\alpha\beta}\right]} = k_{\alpha\beta} \quad (14.20)$$

To summarize, in this chapter we discussed the ME as a model for macrostate dynamics, the probability of a state, and rate coefficients. We defined a number of useful variables and functions such as the MFPT and the committor. The MFPT can be measured experimentally, while the committor helps to understand the process on a more detailed level. We discussed the transition kernel, which is a useful tool to compute these entities.

In the next section, we consider computer simulations with trajectories. Trajectories are the time evolution of phase space points. They provide the most detailed picture of the dynamics (time courses of PS-s). We will focus on determining the rate of transition between MS-s with trajectories.

References

1 McQuarrie, D.A. and Simon, J.D. (1997). *Physical Chemistry: A Molecular Approach*. Sausalito, California: University Science Books.

2 Schutte, C. and Sarich, M. (2013). *Metastability and Markov State Models in Molecular Dynamics: Modeling, Analysis, Algorithmic Approaches*, vol. 24. American Mathematical Society.

3 Weinan, E. and Vanden-Eijnden, E. (2010). Transition-path theory and path-finding algorithms for the study of rare events. *Ann. Rev. Phys. Chem.* 61: 391–420.

4 Ma, P., Cardenas, A.E., Chaughari, M.L., et al. (2017). The impact of protonation on early translocation of anthrax lethal factor: kinetics from molecular dynamics simulations and milestoning theory. *J. Am. Chem. Soc.*, 139: 14837–14840.

5 Mori, H., Fujisaka, H., and Shigematsu, H. (1974). A new expansion of the master equation. 51: 109–122.

6 Dirac, P.A.M. (1999). *The Principles of Quantum Mechanics*. New York: Oxford University Press.

7 Zwanzig, R. (2001). *Nonequilibrium Statistical Mechanics*. Oxford: Oxford University Press.

8 Faradjian, A.K. and Elber, R. (2004). Computing time scales from reaction coordinates by milestoning. *J. Chem. Phys.* 120 (23): 10880–10889.

15

Direct Calculation of Rate Coefficients with Computer Simulations

15.1 Computer Simulations of Trajectories

We consider an exact formulation of the dynamics in classical mechanics. A macrostate α includes many microscopic or precise states (PS) which can be followed individually, and then combined to provide the dynamics of the whole MS. In the most detailed definition, a PS corresponds to a trajectory in phase space. Knowledge of the momenta and coordinates at an initial time, say $(\mathbf{x}_0 = \mathbf{x}(t = 0), \mathbf{p}_0 = \mathbf{p}(t = 0))$, completely determines the full trajectory $(\mathbf{x}(t), \mathbf{p}(t))$ at all times t. The trajectory is obtained by the solution of the Hamilton equations of motion (see Eq. (13.7) for the definition of the Hamiltonian)

$$\dot{\mathbf{x}} = \frac{\partial H}{\partial \mathbf{p}} = \mathbf{M}^{-1}\mathbf{p} \quad \dot{\mathbf{p}} = -\frac{\partial H}{\partial \mathbf{x}} = -\frac{dU(\mathbf{x})}{d\mathbf{x}} \tag{15.1}$$

A variety of algorithms are available to determine the coordinate and the momenta from Eq. (15.1) using a finite time step, Δt. For example, the simplest algorithm of them all, is the Euler's integrator, written below for a single integration step

$$\mathbf{x}(t + \Delta t) = \mathbf{x}(t) + \Delta t \cdot \mathbf{M}^{-1}\mathbf{p}(t)$$

$$\mathbf{p}(t + \Delta t) = \mathbf{p}(t) - \Delta t \cdot \frac{dU(\mathbf{x}(t))}{d\mathbf{x}} \tag{15.2}$$

The integration with a small time-step, Δt, is repeated many times (say N) to achieve the desired time scale (say $N\Delta t$). Moderate time scales are necessary to reproduce the behavior of transitions between PS-s. However, the times that we are typically interested in and for which experimental data are readily available are for transitions between MS-s. We will be particularly interested in calculations of the rate coefficients, $k_{\alpha\beta}$, that require long time trajectories. This is since a significant statistics of transitions between MS-s is needed in the computations of rate coefficients.

A prime use of the numerically constructed trajectories is to compute statistical mechanic averages. We review first equilibrium averages before addressing

Molecular Kinetics in Condensed Phases: Theory, Simulation, and Analysis,
First Edition. Ron Elber, Dmitrii E. Makarov and Henri Orland.
© 2020 John Wiley & Sons Ltd. Published 2020 by John Wiley & Sons Ltd.

questions about kinetics. Consider a quantity of interest, A, which depends on the coordinates and the momenta. We determine the equilibrium average of A, which we denote by $\langle A \rangle_{eq}$ using the equilibrium phase space probability density $P_{eq}(\mathbf{x}, \mathbf{p})$

$$\langle A \rangle_{eq} = \int_\Gamma A(\mathbf{x}, \mathbf{p}) P_{eq}(\mathbf{x}, \mathbf{p}) \cdot d\mathbf{x} \cdot d\mathbf{p} \tag{15.3}$$

where Γ denotes the entire phase space volume (coordinates and momenta). In the micro-canonical ensemble of a constant number of particles N, volume V, and energy E, the equilibrium distribution has the same weight for every configuration with the same energy E.

$$P_{eq}(\mathbf{x}, \mathbf{p}) = \frac{\delta(E - H(\mathbf{x}, \mathbf{p}))}{\int_\Gamma \delta(E - H(\mathbf{x}, \mathbf{p})) d\mathbf{x}\, d\mathbf{p}} \tag{15.4}$$

For the canonical ensemble (constant number of particles N, volume V, and temperature T) we use the Boltzmann distribution (see Eq. (13.7)).

The key for the use of trajectories to compute statistical mechanical averages is the ergodic hypothesis (Eq. 13.3). The ergodic hypothesis states that averages computed over phase space densities are the same as averages over configurations of equilibrium trajectories sampled as a function of time. We say that the system is ergodic if all energetically accessible phase space points are reached in a single trajectory. This sample reproduces the distribution $P_{eq}(\mathbf{x}, \mathbf{p}) d\mathbf{x} d\mathbf{p}$ when the time snapshots are binned into phase space volume elements.

Different algorithms must be used to generate samples from different ensembles. For example, the Hamilton equations of motion (Eq. (15.1)) sample phase space points from the microcanonical ensemble (Eq. (15.4)) and the Langevin Eq. (Chapter 1) from the canonical ensemble. The sampling is accurate and complete if the trajectory is ergodic and the ensemble large enough to yield averages with small statistical errors.

Time averages of systems in *local* equilibrium can be computed by trajectories as well. Consider for example the hierarchy of states shown in Figure 13.2. There is a window of times that are significantly shorter than t_c the typical time for transitions between the MS-s and are much longer than t_l, the time for transitions between PS-s. This time interval can be used for local averaging within a metastable state α. We write for a local equilibrium average within MS α

$$\langle A \rangle_{eq,a}(t) = \lim_{t_l \ll t \ll t_c} \frac{1}{t} \int_{t_0}^{t+t_0} A(\mathbf{x}_\alpha(t'), \mathbf{p}_\alpha(t')) dt' \tag{15.5}$$

The use of the ME to study the MS-s in the previous section is justified for times larger than t_l and of order of or longer than t_c. The separation of time scales (the requirement that $t_c \gg t_l$) is an important tool in the construction of models for kinetics. It is, however, not the only approach and other formulations can be helpful. One should bear in mind that some processes do not show the desired separation. The migration of a molecule on a flat surface will experience many similar timescales as it diffuses spatially.

From the discussion so far, it is obvious that the ergodic hypothesis is critical for the simulations, not only for equilibrium calculations, but also for studies of kinetics. Transitions between MS-s are frequently considered while assuming a local equilibrium at the MS-s. We therefore need algorithms that provide the correct equilibrium distributions, even when we investigate kinetics. It is not obvious that different numerical algorithms to solve the Hamilton equations of motion preserve the ergodic equality (Eq. (13.3)). For the ergodic hypothesis to hold, it is necessary that the statistical equivalence between time and phase space averages is retained. The algorithm must conserve phase space volume as a function of time, i.e. $d\mathbf{x}(t)d\mathbf{p}(t) = $ constant, which is the Liouville theorem. [1] The same equality must hold even if the time step, Δt, is finite as in numerical calculations.

To conserve the volume in phase space we require that the Jacobian of the transformation from $d\mathbf{x}(t)d\mathbf{p}(t)$ to $d\mathbf{x}(t + \Delta t)d\mathbf{p}(t + \Delta t)$ is one.

The Jacobian is given by the following determinant

$$J(\mathbf{x}(t)\mathbf{p}(t) \rightarrow \mathbf{x}(t + \Delta t)\mathbf{p}(t + \Delta t)) \equiv J(\mathbf{x}\mathbf{p} \rightarrow \mathbf{x}'\mathbf{p}')$$

$$J(\mathbf{x}\mathbf{p} \rightarrow \mathbf{x}'\mathbf{p}') = \begin{vmatrix} \dfrac{\partial \mathbf{x}'}{\partial \mathbf{x}} & \dfrac{\partial \mathbf{x}'}{\partial \mathbf{p}} \\ \dfrac{\partial \mathbf{p}'}{\partial \mathbf{x}} & \dfrac{\partial \mathbf{p}'}{\partial \mathbf{p}} \end{vmatrix} = 1 \tag{15.6}$$

Setting the Jacobian to one, is one realization of the Liouville theorem [1]. Eq. (15.6) offers a useful and straightforward test for different algorithms. Consider for example the Euler's integrator (Eq. (15.2)), we write the Jacobian for a one-dimensional system

$$J = \begin{vmatrix} \dfrac{\partial x(t + \Delta t)}{\partial x(t)} & \dfrac{\partial x(t + \Delta t)}{\partial p(t)} \\ \dfrac{\partial p(t + \Delta t)}{\partial x(t)} & \dfrac{\partial p(t + \Delta t)}{\partial p(t)} \end{vmatrix} =$$

$$= \begin{vmatrix} 1 & \Delta t \cdot M^{-1} \\ -\Delta t \dfrac{d^2 U(x(t))}{dx^2} & 1 \end{vmatrix} = 1 + \Delta t^2 M^{-1} \dfrac{d^2 U(x(t))}{dx^2} \tag{15.7}$$

Hence, $J \neq 1$ for the general case and the Euler's trajectories violate the Liouville theorem. The averages computed with Euler's sampling would not provide the desired equilibrium distribution of the microcanonical ensemble. If the system is to stay in a neighborhood of a local minimum (and the second derivative of the potential energy is positive) the errors are positive and accumulate in time. Positive curvatures are frequently the norm since molecular trajectories at room temperature spend significant time near energy minima. Long trajectories will provide particularly bad statistical weights for the phase space points sampled.

One of the most popular molecular dynamics integrators is the so-called Verlet's algorithm [2] which was already discussed by Newton. [3] There are several implementations of the Verlet approach, and here we consider the velocity or the momentum Verlet.

$$\mathbf{x}_{i+1} = \mathbf{x}_i + \mathbf{M}^{-1}\mathbf{p}_i \cdot \Delta t - \frac{1}{2}\Delta t^2 \mathbf{M}^{-1}\nabla U(\mathbf{x}_i)$$

$$\mathbf{p}_{i+1} = \mathbf{p}_i - \frac{\Delta t}{2}(\nabla U(\mathbf{x}_i) + \nabla U(\mathbf{x}_{i+1})) \tag{15.8}$$

The index i denotes the discrete time. The algorithm seems deceptively simple and with local errors proportional to Δt^3 which is not impressive. However, the conservation of the phase space volume element is of paramount importance. We leave it as an exercise to the reader to prove that the Verlet algorithm preserved the phase space volume for an arbitrary time step using the procedure outlined in Eq. (15.7). It is, therefore, a sound choice for an algorithm to conduct statistical mechanic averages counting on the ergodic hypothesis.

Besides the phase space volume conservation, the existence of a "shadow" Hamiltonian in the Verlet algorithm adds significantly to its stability. For any finite time step, Δt, there exists a Hamiltonian from which the Verlet algorithm (Eq. (15.8)) can be derived exactly, using Eq. (15.1). [4] We can write the shadow Hamiltonian as a power series of Δt, but in general we do not know its full functional form. Nevertheless, the mere existence of the shadow Hamiltonian is already useful. If an approximate energy is conserved, it is likely to provide bounds on the exact energy.

Hamiltonian dynamics provides phase space points sampled from well-defined statistics (the microcanonical ensemble). It provides a highly detailed picture of the system, accounting for all atoms using classical mechanics. Consider a process in aqueous solution, which we describe with Hamilton dynamics. Our prime interest is usually in the solute, e.g. a butene molecule undergoing a transition from cis to trans state. To describe the solute accurately we are required to simulate the dynamics of each of the solvent components that interacts with it: water molecules, ions, osmolytes, and other solutes. The number of solvent molecules can be large (simulations of millions of solvent molecules are not unheard of) adding significantly to the computational complexity.

A simplifying consideration is based on timescale separation. The solvent motions are frequently rapid compared to the solute internal dynamics. They can be thought of as the transitions between PS-s that are collected to form an MS, which is an average over solvent motions. Many phenomenological approaches build on this observation and approximate and simplify the description of the solvent. Langevin dynamics offers such a model. It reduces the computational complexity by adding a friction, $-\gamma\dot{\mathbf{x}}$, and a random force, ξ, to model solvent effects and remove the explicit description of solvent PS-s. The random force is sampled from the normal distribution with a zero mean and correlation $\langle\xi(0)\xi(t)\rangle = 2\gamma k_B T\delta(t)$ (see Chapter 1. Eq. (1.12) and (1.15)).

In this picture, the solvent operates on the solute in two modes. First, it provides thermal energy to the solute by collisions with the solvent molecules. The collisions are modeled by a noise term, the random force ξ. Second, the solvent takes away energy from the solute, to balance the energy given in the first mode, such that the temperature of the solute remains constant. A friction force, $-\gamma\dot{\mathbf{x}}$, models the second effect of the solvent. The equations that capture both contributions are variants of the Langevin Eq. (see also Chapter 1 for extensive discussion of the Langevin equation).

$$\mathbf{M}\ddot{\mathbf{x}} + \gamma\dot{\mathbf{x}} + \frac{dU(\mathbf{x})}{d\mathbf{x}} - \xi = 0$$

$$\gamma\dot{\mathbf{x}} + \frac{dU(\mathbf{x})}{d\mathbf{x}} - \xi = 0 \qquad (15.9)$$

The lower equation is for overdamped Langevin dynamics for time scales longer than the time of velocity relaxation to equilibrium. The acceleration, which is the change of the velocity as a function of time is averaged to zero on times longer than the velocity relaxation time. In that case we ignore the inertial term, $\mathbf{M}\ddot{\mathbf{x}}$ (Chapter 1). The noise term balances the frictional dissipation, $-\gamma\dot{\mathbf{x}}$, to retain a constant temperature, T. [5] Eq. (15.9) generates a set of coordinates that are sampled from the canonical ensemble (Eq. (13.7)), augmenting the Hamilton's equations of motions that generate phase space points from the microcanonical ensemble.

Another useful computational feature of Langevin dynamics is that the trajectories are more likely to satisfy the ergodic hypothesis. Hamiltonian dynamics in low dimensions is frequently trapped in a subspace of the energetically accessible PS-s and is not ergodic, i.e. the time average is not equivalent to phase space average. For example, the Hamiltonian trajectory may lack sufficient energy to cross barriers between MS-s. In contrast, the random force of the Langevin dynamics makes it possible for the trajectories to pass essentially any finite barrier height given sufficient time. Hence, ergodicity is easier to satisfy in Langevin dynamics compared to a solution of the Hamilton equations. As a result, the Langevin Eq. is more popular in low dimensional model systems that are easier to analyze theoretically using stochastic and ergodic dynamics.

Because of the noise term, which is not a continuous function of time, an integrator of Eq. (15.9) requires the use of stochastic calculus. [6] For more details on integrating stochastic differential equation see Chapter 1. We list below a high-quality integrator by Leimkuhler and Matthews [4], which is called BAOAB. Eq. (15.10) provides a concrete split and order of integration steps that we denote by (A), (B) and (O).

$$\Delta \mathbf{x} = \mathbf{M}^{-1}\mathbf{p} \cdot \Delta t \qquad \text{(A)}$$

$$\Delta \mathbf{p} = -\nabla U(\mathbf{x}) \cdot \Delta t \qquad \text{(B)} \qquad\qquad (15.10)$$

$$\Delta \mathbf{p}' = \left[\int_0^{\Delta t} -\gamma \mathbf{p} + \sqrt{2\gamma k_B T} M^{1/2}\xi\right]dt \text{ (O)}$$

The three steps (A), (B) and (O) are used in the order suggested by the name BAOAB. Integrate (B) first with a small and finite time step, Δt, then integrate step (A), (O), and (A) and (B) again.

(A) and (B) are deterministic steps and can be integrated using Euler's algorithm. Step (O) is more difficult to integrate since it includes the random force. Nevertheless, it has an analytical solution [4] provided in Eq. (15.11) which we exploit

$$\text{(O):} \quad \mathbf{p}(t) = \exp(-\gamma t)\mathbf{p}(0) + \sqrt{k_B T} M^{1/2}\sqrt{1 - \exp(-2\gamma t)}\xi(t) \qquad (15.11)$$

It is somewhat disconcerting that there is more than one model for the dynamics of the system (Hamiltonian or Langevin) without a clear bridge between them. To those who are concerned we note that there is such a bridge that takes us from a full Hamiltonian description of a large system to stochastic dynamics with a smaller number of variables. The resulting equation is called the Generalized Langevin Equation (GLE)

$$\frac{d\mathbf{p}}{dt} = -\int_0^t \Gamma(t - t')\mathbf{p}(t')dt' + \sqrt{2\gamma k_B T} M^{1/2}\xi(t) \qquad\qquad (15.12)$$

The formulation depends on a construction of a time-dependent "friction", $\Gamma(t - t')$, and a corresponding random force. Eq. (15.12) is derived from Hamiltonian dynamics with a number of degrees of freedom much larger than the number of variables that are considered explicitly in Eq. (15.12) (see Chapter 9). The randomness emerges by averaging over the initial phase space points of degrees of freedom of lesser interest. For example, we average over the coordinates and the velocities of the solvent molecules to retain explicit equations for only the solute degrees of freedom. The Langevin Eq. is obtained from Eq. (15.12) if we set $\Gamma(t) = \gamma \cdot \delta(t)$, which can be interpreted as solvent molecules that respond to solute motions infinitely fast. However, since the response time of the solvent molecules is finite and it is not always fast in reference to solute internal motions, a time dependent memory kernel is obtained.

Eq. (15.12) is difficult to integrate numerically. Every time step forward $\mathbf{p}(t + \Delta t) \cong \mathbf{p}(t) + \int_t^{t+\Delta t} \frac{d\mathbf{p}}{dt} \cdot dt$ includes on the right hand side an integration over the trajectory history, the range of the memory depends on how rapidly the friction function $\Gamma(t)$ decays to zero as a function of the time t. It is also challenging to determine and compute the functional form of the time-dependent friction from first principle Hamiltonian dynamics. These are perhaps the reasons why the GLE is not widely used in molecular simulations.

There is an analogy between the ME and GME and the LE and GLE (Langevin Equation and Generalized Langevin Equation). In both cases, the exact formulation is leading to a kernel or a memory function, which is time-dependent. Simplified and widely used versions are constructed by setting the friction or the rate coefficient to have very short memories, respectively: $\Gamma(t) \to \gamma \cdot \delta(t)$ and $k(t) \to k \cdot \delta(t)$ where $\delta(t)$ is the Dirac's delta function.

The next question we address is how we connect the results of trajectory calculations to the rate theory we discussed in Chapter 8.

15.2 Calculating Rate with Trajectories

The algorithms discussed in 15.A generate trajectories in coordinate or phase space as a function of time. Given initial conditions, the trajectories can be unique (deterministic) or sampled from a distribution of pathways (stochastic). Regardless of the nature of the trajectories we ask what is the probability that a trajectory will start and end at given points or states. More precisely, to compute transition probabilities from state α to state β we determine first the probability to sample a particular PS in state α - $P_{\alpha, LE}(\mathbf{x}_\alpha, \mathbf{p}_\alpha)$ assuming local equilibrium in α. Given the initial coordinate and momentum vectors $(\mathbf{x}_\alpha, \mathbf{p}_\alpha)$ in MS α we compute a trajectory with the dynamics of choice. Finally, we determine if the trajectory ends at an absorbing and terminating MS β at a desired time t (or not).

We consider, both, deterministic and stochastic dynamics. The probability density of ending at the phase space point $(\mathbf{x}_\beta, \mathbf{p}_\beta)$ at time t and that the trajectory starts at time zero at $(\mathbf{x}_\alpha, \mathbf{p}_\alpha)$ is

$$P(\mathbf{x}_\beta \mathbf{p}_\beta, t; \mathbf{x}_\alpha \mathbf{p}_\alpha, 0) = P(\mathbf{x}_\beta \mathbf{p}_\beta, t \mid \mathbf{x}_\alpha \mathbf{p}_\alpha, 0) \delta((\mathbf{x}_\alpha \mathbf{p}_\alpha) - (\mathbf{x}(0)\mathbf{p}(0))) \quad (15.13)$$

where $P(y \mid x)$ is a conditional probability. Given the event x, it is the probability to observe y. The joint probability is $P(y; x)$, which is the probability of observing both events.

For a deterministic trajectory we can write more explicitly the joint probability:

$$P(\mathbf{x}_\beta \mathbf{p}_\beta, t; \mathbf{x}_\alpha \mathbf{p}_\alpha, 0) = \delta((\mathbf{x}_\beta \mathbf{p}_\beta) - (\mathbf{x}(t)\mathbf{p}(t))) \delta((\mathbf{x}_\alpha \mathbf{p}_\alpha) - (\mathbf{x}(0)\mathbf{p}(0)))$$

To obtain the probability of transition from MS α to MS β we sum over all the PS states in MS α, which are the initial conditions for the trajectories, and the phase space points at the boundary of the absorbing MS β, we have

$$P(\beta, t; \alpha, 0) = \frac{1}{N} \sum_{\mathbf{x}(0), \mathbf{p}(0)} \int dx_\alpha dp_\alpha dx_{\beta B} dp_{\beta B} P(\mathbf{x}_{\beta B} \mathbf{p}_{\beta B}, t \mid \mathbf{x}_\alpha \mathbf{p}_\alpha, 0) \cdot$$
$$\cdot P_\alpha(\mathbf{x}_\alpha \mathbf{p}_\alpha) \cdot \delta((\mathbf{x}_\alpha \mathbf{p}_\alpha) - (\mathbf{x}(0)\mathbf{p}(0))) \tag{15.14}$$

Where $(dx_{\beta B} dp_{\beta B})$ denotes an integration element on the absorbing boundary of the entry to state β, N is the number of trajectories and the summation is over all the initial conditions of the trajectories. Eq. (15.14) describes a sequence of steps to prepare and conduct trajectory calculations: (i) Sample a phase space point in MS α according to the known probability $P_\alpha(\mathbf{x}_\alpha, \mathbf{p}_\alpha)$, (ii) Integrate the equations of motion to time t, (e.g. Eq. (15.8)) to obtain $(\mathbf{x}(t)\mathbf{p}(t))$. (iii) Check if the phase space point at time t is on the boundary of MS β. If it is, add its contribution to the integral and terminate that trajectory. If not, do not add the trajectory contribution to the integral and retain the trajectory for future reference. For example, it may cross the terminating boundaries at later or earlier times. Return to (i).

If α is a metastable state, we can derive a relationship between $P(\beta, t; \alpha, 0)$ and $w_{\beta\alpha}(\mathbf{x}_{\beta B} \mathbf{p}_{\beta B}, \mathbf{x}_\alpha, \mathbf{p}_\alpha, t)$ (section 14.4) given that state β is directly accessible from α. We remind the reader that the kernel $w_{\beta\alpha}(\mathbf{x}_{\beta B} \mathbf{p}_{\beta B}, \mathbf{x}_\alpha, \mathbf{p}_\alpha, t)$ is the probability to transition to β at time t at phase space point $(\mathbf{x}_{\beta B} \mathbf{p}_{\beta B})$ given that at time zero the system was in state α at $(\mathbf{x}_\alpha, \mathbf{p}_\alpha)$.

$$P(\beta, t; \alpha, 0) = \int w_{\alpha\beta}(\mathbf{x}_{\beta B}, \mathbf{p}_{\beta B}, \mathbf{x}_\alpha, \mathbf{p}_\alpha, t) P_{\alpha, LE}(\mathbf{x}_\alpha, \mathbf{p}_\alpha, 0) \cdot dx_\alpha \cdot dp_\alpha$$
$$\cdot dx_{\beta B} \cdot dp_{\beta B} \tag{15.15}$$

If the relaxation within the state is such that the local equilibrium is always maintained we recover the ME and use the time dependent solution of the kernel (Eq. 14.5) to write

$$\log[P(\beta, t; \alpha, 0)] = \log(k_{\alpha\beta} P(\alpha, 0)) - \sum_\gamma k_{\alpha\gamma} t \tag{15.16}$$

From the slope of $\log[P(\beta, t; \alpha, 0)]$ as a function of time we can extract the total rate coefficient of state $\alpha - \sum_\gamma k_{\alpha\gamma}$ but not the individual rate $k_{\alpha\beta}$. If the system includes only two states then the overall rate coefficient is reduced to $k_{\alpha\beta}$.

The MFPT can be calculated from the information provided by Eq. (15.14). Consider the definition of the MFPT. It is the average time of trajectories initiated at state α that hit state β for the first time. We write yet another expression for the MFPT in addition to Eq. (13.6) and Eq. (14.12).

$$\langle \tau \rangle = \frac{\int_0^\infty t P(\beta, t; \alpha, 0) dt}{\int_0^\infty P(\beta, t; \alpha, 0) dt} \tag{15.17}$$

With the kernel of the ME at hand we obtain an explicit expression for the MFPT

$$\langle \tau \rangle = \frac{\int_0^\infty t P(\beta, t; \alpha, 0) dt}{\int_0^\infty P(\beta, t; \alpha, 0) dt} = \frac{\int_0^\infty t \cdot k_{\alpha\beta} P(\alpha, 0) \exp\left[-\sum_\gamma k_{\alpha\gamma} t\right] dt}{\int_0^\infty k_{\alpha\beta} P(\alpha, 0) \exp\left[-\sum_\gamma k_{\alpha\gamma} t\right] dt} = \frac{1}{\sum_\gamma k_{\alpha\gamma}}$$

(15.18)

Again, the individual rate coefficient, $k_{\alpha\beta}$, can be extracted using the above average if the system has only two MS-s, or if the other states are blocked or removed from considerations. For example, consider all the trajectories initiated at α, remove all the trajectories that end up in a state different from β. The remaining trajectories between the two states (α and β) follow two-state kinetics.

The exponential decay of the population that we use in equations (15.16) and (15.18) is not guaranteed. More complex time-dependent probabilities are possible. It is therefore useful to report the time course or moments of the distribution of the first passage time, if the statistics is sufficient. Note that it is straightforward to compute the MFPT for different distributions (Eq. 15.18 and 13.6); however, the calculations of a rate coefficient that require on exponential decay in time are not always possible.

References

1 McQuarrie, D. (2000). *Statistical Mechanics*. Sausalito: University Science Books.
2 Verlet, L. (1967). Computer "experiments" on classical fluids. I. Thermodynamical properties of Lennard-Jones molecules. *Phys. Rev.* 159 (1): 98–103.
3 Hairer, E., Lubich, C., and Wanner, G. (2003). *Acta Numer.* 12: 399–450.
4 Leimkuhler, B. and Matthews, C. (2015). *Molecular Dynamics with Deterministic and Stochastic Numerical Methods*, vol. 39. Springer.
5 Kubo, R. (1966). Fluctuation-dissipation theorem. *Rep. Prog. Phys.* 29: 255.
6 Karlin, S. and Talor, H.W. (1975). *A First Course in Stochastic Processes*, 2e. Elsevier.

16

A Simple Numerical Example of Rate Calculations

Below, we illustrate the use of trajectories to compute the thermodynamics and kinetics of a model system. We use the overdamped Langevin Eq. (Eq. 15.9, lower line) to study dynamics on the Mueller potential (Fig 16.1). [1] The Mueller potential is a two-dimensional energy landscape that is frequently considered a model system for the study of activated processes, or processes that overcome significant energy barriers.

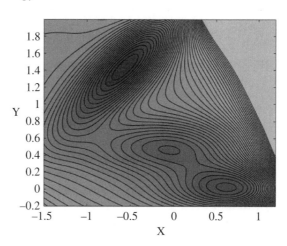

Figure 16.1 A two-dimensional contour plot of the Mueller energy function. [1] The contour lines are of equi-potential values, and the color code varies from blue to yellow, where the blue is the lowest energy. The yellow domain is of high energy and is inaccessible to trajectories at a moderate temperature. (*See color plate section for color representation of this figure*).

The potential is a sum of four terms, each of them is of the form $A \cdot \exp[a(x - x_0)^2 + b(x - x_0)(y - y_0) + c(y - y_0)^2]$, the different parameters are listed below in Table 16.1

The Mueller potential has three minima and two saddle points. The locations of the minima, saddles, and their energies are provided in Table 16.2

The deepest minimum (A) is on the top left corner near x~−0.6. The shallowest minimum (C) is at the center around x~0.0 and the minimum with intermediate depth is at the lowest right corner at x~0.6 (B). The highest barrier

Molecular Kinetics in Condensed Phases: Theory, Simulation, and Analysis,
First Edition. Ron Elber, Dmitrii E. Makarov and Henri Orland.
© 2020 John Wiley & Sons Ltd. Published 2020 by John Wiley & Sons Ltd.

Table 16.1 A list of parameters for the Mueller potential.

	1	2	3	4
A	−200	−100	−170	15
a	−1	−1	−6.5	0.7
b	0	0	11	0.6
c	−10	−10	−6.5	0.7
x_0	1	0	−0.5	−1
y_0	0	0.5	1.5	1

Table 16.2 The locations and energies of the minima and saddle points of the Mueller potential.

	X	Y	U(x,y)
Minimum A	−0.558	1.442	−146.700
Minimum B	0.623	0.028	−108.167
Minimum C	−0.050	0.467	−80.768
Saddle AC	−0.822	0.624	−40.665
Saddle BC	0.212	0.293	−72.249

is between the deepest and the shallowest minima (−40.665−(−146.700) =106.035). The barrier between the shallow and intermediate minima is low (−72.249−(−80.768)) =8.519). It is therefore expected that transitions between the two minima at the lower half of the potential energy plot will be easier to sample during a trajectory with temperature of 10 (see below).

We consider a stochastic trajectory that follows overdamped Langevin dynamics (Eq. (15.9)) and is computed on the Mueller potential. We use a simple algorithm to generate trajectories according to overdamped Langevin dynamics:

$$\mathbf{x}(t + \Delta t) = \mathbf{x}(t) + \frac{\Delta t}{\gamma}(\xi(t) - \nabla U(\mathbf{x}(t))) \qquad (16.1)$$

Interestingly, the random force and the friction term mask the problems with phase space conservation of the Euler's algorithm that we discussed in the context of deterministic dynamics. Eq. (16.1), which is a variant of the Euler's algorithm, is working quite well. We discuss below a concrete numerical example. The random force, $\xi(t)$, is sampled from a normal distribution with a zero mean and variance of $2\gamma k_B T/\Delta t$. The friction coefficient and the Boltzmann constant are set to one. The temperature is 10. Hence the barrier between the shallowest and the intermediate minima is of order of kT ($10 \sim 8.519$). On the other hand, the largest barrier is of order of $10kT$. The initial coordinates $\mathbf{x}(0)$ were at (0.5, 0.0) or at the shallow minimum. The time step was 9×10^{-6} and the equations of motion were integrated for two billion (2×10^9) steps.

It is interesting to examine if the trajectory is long enough to represent an equilibrium state. The ergodic hypothesis (Eq. (15.6)) equates equilibrium average over phase space to averages over time for an infinitely long trajectory. Sampled configurations (at constant time intervals of 2000 steps to reduce correlations between the conformations) are displayed on the Mueller coordinate space in Figure 16.2. The "intensity" of the lines in the major states suggests (qualitatively) that the system spent considerable time at each energy minimum and nowhere else. Out of a total of a million points sampled we find 969 030 configurations in the upper left minimum (which we call $\mathbf{u}$, also the upper minimum A) and only 30 970 configurations in the combined set of the two lower minima (which we call $\mathbf{d}$, the combination of minima B and C). This sampling emphasizes the importance of the state $\mathbf{u}$ that captures about 96.9% of the population at equilibrium. Do the statistics we extracted from the trajectory indeed represent an equilibrium state?

A simple "sanity" check of the convergence of the trajectory is to use only the first half of the sample points and examine if observable values have been changed. If we do that, we find 483 405 configurations in the deeper minimum. This is 96.7% of the population, which is not far from our previous estimate that was based on a larger sample. Hence the check shows consistency between two

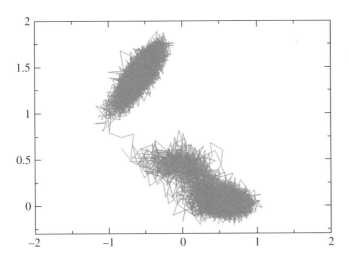

Figure 16.2 A segment of a single, long trajectory (one hundred thousand sample points) on the Mueller potential (see a contour plot of the Mueller potential in Figure 16.1). The minima are sampled frequently, however, transitions from the upper to the lower minima are rarely observed even in the long trajectory.

averages that were conducted with trajectories of different lengths. It, therefore, suggests that an equilibrium state has been reached.

In addition to the study of equilibrium or thermodynamics, we can use a long trajectory to investigate the kinetics of transitions between states. While our previous discussion considered the time evolution of a population or a number of trajectories, it is possible to use a single very long trajectory for the same task.

To extract a statistically meaningful set of transitions from a single trajectory, we chop the trajectory at the instances in which it enters a new state. The collection of trajectory fragments from the point of insertion into a new state (say α) until they "touch" another state (say β) provides an ensemble of events of transitions that can be used to quantify the kinetics and compute $P(\alpha, 0; \beta, t)$. Examples of transitional events during a single long trajectory are shown in Figure 16.3.

Figure 16.3 illustrates two important concepts of kinetics: (i) metastability and (ii) rare events. The Mueller potential has three different energy minima (Figure 16.1 and Table 16.2), so a first attempt is to associate each of the minima with a metastable state. However, as is sketched in Figure 13.2, different energy minima separated by barriers comparable to, or lower than the thermal energy are collected into a single macrostate. By visually inspecting a long trajectory

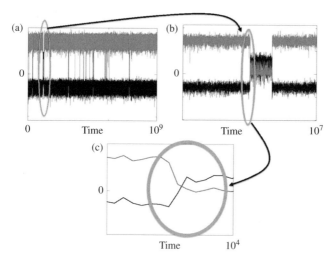

Figure 16.3 Transitions between states in the Mueller potential extracted from a long trajectory are displayed at different magnifications of the time scale. The black and red lines are the trajectory values along the horizontal and vertical axes as a function of time respectively. The lowest energy minimum is populated when the black and red lines are well separated and the black curve is negative. The time window is magnified as we shift from panels a to c. Note the difference in time scales and the blue circles around a transition that we choose to magnify. Similar pictures illustrating metastability can be found elsewhere in the book, see Figure 1.2 and 8.2. (*See color plate section for color representation of this figure*).

(Figure 16.2) we observe that the trajectory oscillates back and forth many times between the two **d** minima. We remind the reader that the barrier separating the shallow minima of 8.519 is close to kT (10) which makes the rapid transitions understandable. Grouping the two lower minima to a single macrostate is consistent with the behavior of the trajectory (Figure 16.2 and 16.3). The lowest energy minimum at the upper left portion of the energy landscape is one macrostate and the two other minima form another single macrostate. Of course, more rigorous approaches than visual inspection can be used to identify macrostates. For example, researchers have used clustering [2], or separation of time scales [3] for this task. Nevertheless, for the example at hand visual inspection of trajectories on the two-dimensional energy landscape works just as well.

The second concept is of a rare event. As is illustrated in Figure 16.3. the transitions between the states are not frequent. Over a period of billions of integration steps, we observed only a few transitions. To obtain meaningful statistics of these events in a single run, a much longer trajectory than typical time interval separating the two transition events is required. Another observation is that the transitions are extremely rapid compared to the time the trajectory dwells or waits in the well. The transition time cannot be estimated from Fig 16.3.a or 16.3.b, but in Figure 16.3.c the trajectory transition time is spread over several data points. A configuration or a data point is saved every 2000 integration steps. Hence, a transition requires a couple of thousands of time steps. In contrast, the dwelling time in the well is millions of time steps, or a thousand times longer than the transition time. The overall rate of the reaction is determined by a sum of both, the transition time, and the dwelling time. Because of the long time that is required for the initiation of an activation, the overall rate of the process is not determined by the actual transition time, but by the waiting time, instead.

We note that in complex systems, the dynamics and kinetics can be considerably more elaborate. We may observe numerous "rare" events of crossing barriers of variable heights and significance. Consider the two energy landscapes sketched below in Figure 16.4

Figure 16.4 A schematic drawing of energy landscapes U(X) as a function of the coordinate X. The solid line illustrates an energy landscape that is likely to have only a single type of activated process and a rare event. The dash line shows an energy landscape with multiple types of crossing of energy barriers with variable heights.

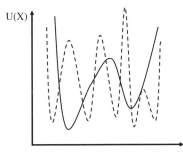

Describing the dynamics and the kinetics on the dashed energy landscape of Figure 16.4 is a significant challenge. Not only we need to describe multiple rare events, but some of the transitions may not be rare and not between states that are in local equilibrium. Approaches that focused on one or a few rare events, may not be appropriate on rough energy landscape with a distribution of barrier heights. We can illustrate these complications on the simple low-dimensionality Mueller's potential as discussed below.

To obtain meaningful statistics for kinetic considerations on the Mueller potential we extracted first passage times from a single long trajectory of 20 billion steps. We saved configurations every 1000 steps and we check for transitions using a two-state model. If the system has the y coordinate larger than 0.75, we call the state an "upper" - **u**. If it is smaller than 0.75, it is a "downer" - **d**. We chop the trajectories into pieces from the first entry to a state until it exits to the other state. We ask, given that the system just entered a state **d** (or **u**) what is the waiting time until a trajectory enters for the first time (passes y=0.75) the other state **u** (or **d**)? The distribution of these waiting times, which are also the first passage times, is shown in Figure 16.5.

Note that in the left panel of Figure 16.5 and in Figure 16.6, the distribution of exit times from the **u** state is close to exponential. The exception is the long tail at times longer than 40 000 that are poorly sampled. In accord with Eq. (14.5) we suggest this state to be metastable. This observation is not surprising since the **u** minimum is the deepest of them all, allowing for ample time for equilibration inside the **u** state before making a transition to the other state. Hence, we expect a separation of time scales between internal and external dynamics.

On the other hand, the **d** state is shallower, making the transition times from the **d** to the **u** state faster and closer to the internal relaxation time within the **d** state. Moreover, the internal relaxation at the **d** state is more complex since it includes transitions between two energy minima that are separated by a barrier. The internal relaxation in **d** is longer than the internal relaxation in **u**.

The first passage time distribution of the **d** state (Figure 16.5, lower panel) is not a simple exponential function in time. According to our test (Eq. 14.2), the relaxation from the **d** state cannot be described by a single MS behavior, and the **d** state is not Markovian in time. Note however that the MFPT of any state is defined as the average time using the distributions shown in Figure 16.5. Hence the MFPT is a meaningful function for both distributions, Markovian or not. What we cannot do is to equate the MFPT with the inverse of a rate coefficient for the **d** state. To make that assignment the FPT distribution needs to be exponential.

A simple approach to extract the rate coefficient from the trajectory data is to plot the logarithm of the probability of the first passage time versus the first passage time. This plot should be a straight line and the negative of the slope the relaxation coefficient, Λ, ($w_{\alpha\beta}(t) \propto \exp(-\Lambda t)$, see Eq. (14.4)). An illustration of the logarithm plot of the relaxation from state **u** is in Figure 16.6.

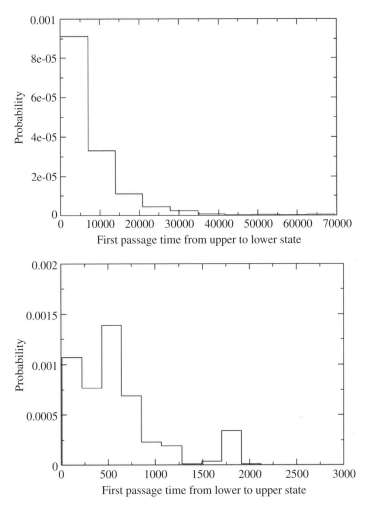

Figure 16.5 The distribution of first passage times (FPT), $\pi(\tau)$, for transitions between the upper and lower states of the Mueller potential (see text for more details). Upper panel: the distribution of FPT for transitions between the **u** state and the **d** state. Lower panel: the FPT distribution for transitions from the **d** state to the **u** state. The times are reported in units of 1000 steps. Hence, a reported time of 1000 corresponds to one million integration steps.

The "microscopic" dynamics we started with, the overdamped Langevin dynamics, is Markovian in the coordinate space of the Mueller system (generation of the next step depends only on one step backward in time, see discussion in Chapter 14). Only after reducing the effective space into two discrete macrostates do we obtain non-Markovian behavior. Hence, memory effects and non-exponential kernels (Eq. (14.15) and (15.12)) are direct results

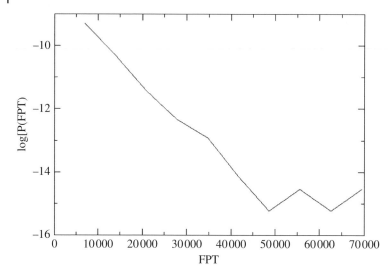

Figure 16.6 The natural logarithm of the probability of the first passage time (FPT) as a function of the FPT. A reasonable straight line is observed for FPT shorter than 50 000. Note also that the longest times are sampled poorly and should not be used in a fit. The coefficient of relaxation rate (the absolute value of the line slope) is $1.4 \ 10^{-4}$ in the time interval $t \in [7{,}000 - 50 \ 000]$.

of coarse-graining microscopic dynamics, and violation of the local equilibrium assumption. Attempting to describe a Markovian system with a large number of degrees of freedom with another system with a (vastly) smaller number of degrees of freedom can lead to long local equilibration periods that compete with transition times between states, and to non-Markovian behavior.

Another consideration is the cost of the calculations. The total length of the trajectory for the analysis of the kinetics was of 20 billion steps. While simulations of this length are easy to conduct for the Mueller potential, they are considerably more challenging for large molecular systems when the computations of each step are more expensive.

A typical measure of the computational complexity of Molecular Dynamics (MD) simulations is the number of force evaluations. Benchmarks of (serial) MD programs suggest that force evaluations (e.g. the calculation of $-\nabla U$ in Eq. 15.8 or step B of Eq. 15.10) require between 90 to 99% of the computational time. A typical time step in simulations of condensed phase systems is a femtosecond or 10^{-15} s. The fastest motions in the system determine the magnitude of the time step (e.g. bond vibrations or rapid collision events). It cannot be increased significantly while retaining an accurate solution of the trajectory. Each step requires at least one force calculation. Twenty billion steps (the length of the Mueller trajectory) will give a physical time of $\sim 2 \times 10^{-5}$ s. This time is far too short to study many biophysical processes, such as enzymatic reactions, membrane transport, and more. These processes have typical time scales of milliseconds (10^{-3} s).

Another major consideration is the clock time (the time in the real world). A single force evaluation on a fast computer requires about a microsecond of real time. A single millisecond event requires 10^{12} steps and million seconds or 11 days of clock time. Technologies different from conventional MD are desired to investigate rare events efficiently. Even more challenging is a sequence of rare events separated by barriers with a wide distribution of heights (e.g. Figure 16.4). These events may or may not be broken to a series of transitions between MS-s that are consistent with a Markovian process.

References

1 Muller, K. and Brown, L.D. (1979). Location of saddle points and minimum energy paths by a constrained simplex optimization procedure. *Theor. Chim. Acta* 53 (1): 75–93.

2 Bowman, G.R. and Pande, V.S. (2014). *An Introduction to Markov State Models and Their Applications to Long Timescale Molecular Simulations.* Springer, p. 139.

3 Shalloway, D. (1996). Macrostates of classical stochastic systems. *J. Chem. Phys.* 105 (22): 9986–10007.

17

Rare Events and Reaction Coordinates

As argued at the end of the last section, computations of long-time trajectories are expensive as measured by the number of required force evaluations and/or computer cycles. The cost of long trajectories is the most significant limitation of MD simulations. This chapter and the one that follows outline calculations of kinetics without the calculations of complete trajectories from reactants to products while retaining an atomically detailed description of the system.

In the present section, we consider the calculations of reaction paths and reaction space. We focus on the calculations of reaction coordinates (RC) since they lead to the most straightforward calculations of rates. There are two general approaches to determine reaction coordinates. One approach is based on a static analysis of the energy landscape while the second technique analyzes trajectory data or stochastic models (see also Chapter 7). The second approach requires that we will have at hand long trajectories before the calculation of the RC can proceed. Since we consider cases in which we are unable to conduct straightforward trajectories to the desired length we will focus on the first approach that does not require such trajectories.

Let $\mathbf{y}$ be a set of coarse variables that forms a reduced space sufficient to describe the reaction progress. Let $\mathbf{y}_a$ and $\mathbf{y}_b$ be the centers of two MS-s on the free energy $F(\mathbf{y})$ that we consider to be reactant and product states. It is an effective energy surface with a smaller number of degrees of freedom. We use the canonical distribution to average over all other degrees of freedom that are not $\mathbf{y}$. In other words, all the degrees of freedom, which are not $\mathbf{y}$, are in local equilibrium. This is a similar assumption to metastability. However, since the $\mathbf{y}$ are continuous variables and the MS-s are discrete states, the time-scale separation of the $\mathbf{y}$ variables is a stronger assumption.

$$F(\mathbf{y}) = -\beta^{-1} \log \left[\int d\mathbf{x} \cdot \delta(\mathbf{y} - \mathbf{y}_0(\mathbf{x})) \exp(-\beta U(\mathbf{x})) \right] \tag{17.1}$$

where $\mathbf{y}_0(\mathbf{x})$ is the mapping from the full conformational space $\mathbf{x}$ to the reduced space of $\mathbf{y}$. Here we assume that the coordinates $\mathbf{y}$ are Cartesian. Otherwise Jacobian must be added for the transformation between the different sets of

Molecular Kinetics in Condensed Phases: Theory, Simulation, and Analysis,
First Edition. Ron Elber, Dmitrii E. Makarov and Henri Orland.
© 2020 John Wiley & Sons Ltd. Published 2020 by John Wiley & Sons Ltd.

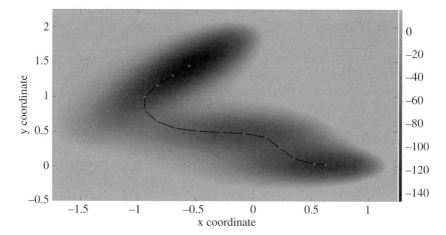

Figure 17.1 A minimum energy path computed by the locally updated planes approach described in reference [3] and Eq. (17.3) from the top left minimum of the Mueller potential to the lower right minimum. The reaction coordinate is presented by a set of discrete configurations along the curve (blue dots). The connecting dashed line is to guide the eye. The colors code the energy value with the red the lowest energies. An energy scale is provided at the right bar. (*See color plate section for color representation of this figure*).

coordinates. For a complete discussion see [1]. In the present description we do not differentiate between optimal pathways on the energy $U(\mathbf{x})$ or the effective (free) energy landscape $F(\mathbf{y})$.

The RC is a continuous one-dimensional object that guides us between the initial macrostate (the reactant coordinate $\mathbf{y}_a$) and the final macrostate (the product coordinates $\mathbf{y}_b$) (Figure 17.1).

The RC has two main realizations. First, it is a one-dimensional, continuous curve connecting the centers of the macrostates. Second, it is a set of non-crossing hypersurfaces progressing from the reactant to the boundaries of the product state. If the number of degrees of freedom of the system is N, the hypersurfaces are of dimension $N - 1$. The pictures of a curve and of hypersurfaces are, of course, connected. For example, it is common to model the hypersurfaces as planes orthogonal to the path at the points in which they cross the curve. The planes are approximations to the hypersurfaces since they cross each other at large distances from the curve; if the curve is not a straight line. The crossing causes an ambiguity in the mapping of a coordinate value to a location at the RC. A point at the crossing of two hyperplanes is mapped to two positions along the RC. Points just before or after plane crossing may be mapped to widely different RC coordinates. The use of planes is therefore restricted to trajectories that remain in the close neighborhood of the curve.

The reaction coordinate is expected to summarize and capture the dynamics of the system relevant to the transition. The type of dynamics can vary widely.

It can be, for example, Hamiltonian and energy conserving, or overdamped Langevin dynamics at a constant temperature. Clearly, summarizing all types of dynamics with a single reaction coordinate is questionable if at all possible. Nevertheless, molecular trajectories of arbitrary dynamics frequently visit the same restricted space. A large portion of the phase or configuration space is not sampled during the reaction progress since the corresponding configurations are of high energy. This observation suggests that the description can be simplified to explicitly include only domains that are likely to be visited during the reaction progress. The promise of the reaction coordinate approach is based on the last observation of limited accessible space. The identification of phase space domains that are important for the progress of the reaction is, therefore, an active research topic.

Consider the configuration space shown in Figure 16.1. The domain highlighted in yellow is of high energy and is not likely to be visited at the moderate temperature of the simulations. The accessible configuration space can be reduced further upon examination of exploratory trajectories, for example, the trajectory in Figure 16.2. We realize that the reactive trajectory spends a considerable length of time at the minima and rarely makes a transition between states **u** and **d**. If they do happen, the rare transitions occur at the neighborhood of a saddle point connecting the minima. A reduced model may include the space of the three minima and a single transition domain between the upper and lower states. Hence, there is a clear bias for a reactive trajectory to pass at the lowest possible energy domains that still keep the space connected.

We build on the above qualitative arguments to define the RC more precisely: A reaction coordinate, $\Phi(l)$, charts a path with the lowest (free) energy barrier between the two minima, $\mathbf{x}_R$ and $\mathbf{x}_P$ (Figure 17.1). For simplicity, we use below only the energy notation. We parameterize the RC curve by $\Phi(l)$ where $l \in [0, 1]$ and Φ is a coordinate vector in the full space, $\mathbf{x}$. The parameter l is zero at the reactant, and one at the product.

The path we consider is called the Steepest Descent Path (SDP). The maximum energy point along the path is the minimal barrier that we must cross on our way from the reactant to the product. Alternatively, we require that each point along the path is energy minimized in all directions with one exception, the direction determined by the local slope of the path ($\mathbf{e}_l \equiv d\Phi/dl$). A differential equation leading asymptotically to the SDP is

$$\frac{d\Phi(l, \tau)}{d\tau} = -\left[\mathbf{I} - \frac{\mathbf{e}_l \mathbf{e}_l^t}{|\mathbf{e}_l^t||\mathbf{e}_l|} \right] \nabla U(\Phi(l, \tau))$$

$$\mathbf{e}_l = \frac{d\Phi(l, \tau)}{dl} \tag{17.2}$$

with the boundary condition $\Phi(0, \tau) = \mathbf{x}_R$ and $\Phi(1, \tau) = \mathbf{x}_P$. Eq. (17.2) is a differential equation for the curve $\Phi(l, \tau)$. We added the fictitious time, τ, as another path variable. At a fictitious time of zero, the path, $\Phi(l, 0)$, is the initial guess to the path that is provided by the user and must satisfy the boundary conditions.

At long times and as $\tau \to \infty$ the path reaches a fixed point, which is the reaction coordinate or the steepest descent path that we seek. The right-hand side of Eq. (17.2) includes the potential gradient multiplying an operator in square brackets. The operator is the identity ($\mathbf{I}$) minus a projection operator along the path direction. The subtraction provides a filtering mechanism that allows moves in any direction with the exception of the current direction of the path $-d\Phi/dl$.

The "long-time" solution is computed with an overdamped Langevin dynamics of a path at zero temperature. There is no noise term in Eq. (17.2) (see Chapter 1 and Eq. (15.9) for Langevin dynamics at a finite temperature) leading asymptotically to a fully quenched and minimized curve between the two end points of the reactants and the products.

In practical calculations, the curve is discretized and solved as a set of discrete points $\{\Phi_i\}_{i=0}^{L}$ along the path with $i = 0$ the reactant and $i = L$ the product.[28] The path slope is estimated as $\mathbf{e}_i = (\Phi_{i+1} - \Phi_{i-1})/|\Phi_{i+1} - \Phi_i|$ where $|...|$ is the norm of a vector. There is no need to estimate the path slope at the first and last points since these points are fixed.

$$\frac{d\Phi_i(\tau)}{d\tau} = -[\mathbf{I} - \mathbf{e}_i \mathbf{e}_i^t]\nabla U(\Phi_i(\tau)) \quad i = 1, \ldots, L-1 \tag{17.3}$$

As written, there is no information in Eq. (17.3) on the motion along the curve l. Eq. (17.3) controls only the dynamics in the direction perpendicular to the curve. For a uniform spatial resolution, which is the optimal choice without prior knowledge of the final answer and the overall path curvature, the grid points of any intermediate path should be equally spaced along the curve. If a significant number of the points are compressed into one segment of the path, other path segments are described poorly. Further refinement of the distribution of the configurations along the path may require a more significant density of points at highly curved segments of the reaction pathway. The first implementation of Eq. (17.3), dubbed Locally Updated Planes, did not consider displacements along the curve. [3] If the distribution of points in the initial guess mapped well to the distribution along the SDP then Eq. (17.3) provides useful discretized pathways. In other cases, however, it is beneficial to equally redistribute the configurations along the curves and allow for displacements along the RC. This can be done by adding a penalty function on the distances between the points along the path.

We define $\Delta l_{i,i+1} = |\Phi_{i+1} - \Phi_i|$. We then add an additional restraint to Eq. (17.3) $R = \sum_{i=1}^{L-1} k(\Delta l_{i,i+1} - \langle \Delta l_{i,i+1} \rangle_L)^2$ where $\langle ... \rangle_L$ is an average over all the L distances between nearby configurations along the path. i.e.,

$\langle \Delta l_{i,i+1} \rangle_L = \frac{1}{L} \sum_{i=0}^{L-1} \Delta l_{i,i+1}$ and k is an empirical constant that is used to penalize deviation from the average. [4] Eq. (17.3) is therefore adjusted to

$$\frac{d\Phi_i(\tau)}{d\tau} = -[\mathbf{I} - \mathbf{e}_i \mathbf{e}_i^t]\nabla_i U(\Phi_i(\tau)) - \nabla_i R \quad i = 1, \dots, L-1 \tag{17.4}$$

Instead of a restraint, it is possible to enforce a uniform distribution of points along the path using Lagrange multipliers, which is the approach used in the string method. [5] Actually, there are a number of methods that build on one form or another of Eq. (17.4). This approach started at reference [3] introducing the Locally Updated Planes (LUP) approach, continuing with the Nudged Elastic Band (NEB) [6] and variants, and concluding with the more recent string approach. [5]

Another class of techniques to compute reaction coordinates is based on optimization of functionals (see also Chapter 7). We define $S[\Phi(l)]$ to be a functional of the path. That is, given a path $\Phi(l)$ the functional assigns a score to it. The reaction coordinate is obtained when the score is minimal. In particular we consider the functional

$$S[\Phi(l)] = \int_{x_R}^{x_P} \sqrt{\nabla U^t \nabla U} \cdot dl \tag{17.5}$$

where we use the same definition for reactant and product as in Eq. (17.2) and dl is an arc-length element. A minimum of this functional with respect to the path provides the SDP. This is illustrated as follows: Compare the functional of Eq. (17.5) to the integral, I, below:

$$I = \int_{x_R}^{x_P} \left| \nabla U^t \vec{dl} \right| \tag{17.6}$$

The scalar product in the integral of Eq. (17.6), $|\nabla U^t \vec{dl}|$, is always smaller or equal than the product of Eq. (17.5) $\left(\sqrt{\nabla U^t \nabla U} \cdot dl \equiv |\nabla U| \cdot |\vec{dl}| \right)$. The functional S is equal to I if the gradient of the potential is parallel or anti-parallel to the path segment $\left(\nabla U \| \vec{dl} \right)$, which is also the minimum of S. A *definition* of the SDP is that the vectors ∇U and the path element $\vec{dl}$ are either parallel or anti-parallel to each other at each point along the path. Hence, the minimization of the functional of Eq. (17.5) provides us with the path we seek. The minimization can be conducted in a discrete form similar in spirit to Eq. (17.4).

$$S' \left[\{\Phi_i\}_{i=0}^L \right] = \frac{1}{2} \sum_{i=1}^{L-1} \sqrt{\nabla_i U^t(\Phi_i)\nabla_i U(\Phi_i)}[\Delta l_{i-1,i} + \Delta l_{i,i+1}] + R \tag{17.7}$$

where we used the trapezoid rule to approximate the line integral of Eq. (17.5). Eq. (17.7) describes a function S' of the whole discrete path. The minimization can be conducted by quenching the whole path with an "energy" given by S' as a function of the fictitious time τ

$$\frac{d\Phi_i}{d\tau} = -\nabla_i S' \qquad i = 1, \ldots, L - 1 \tag{17.8}$$

Similar to the LUP algorithm [3], variants of the functional approach are also possible. An example of a reaction coordinate computed with an adjusted functional approach for the Mueller potential is found in reference. [2]

The advantage of functionals for computing minimum energy paths is the path–score that they provide. The score can be used to accept or reject an attempt of path adjustment; an adjustment which can be large. Given a path, $\Phi(l)$, it is possible to introduce a large perturbation to the path, $\Delta(l)$, and to check the score S' of the new path. If the score of the new path, $\Phi(l) + \Delta(l)$, is lower than the score of $\Phi(l)$ we accept the new path. There is no such global quality measure for paths generated according to Eq. (17.4) and therefore path displacements in the Locally Updated Plane approach are kept small.

The disadvantage of the functional approach is that the calculation of the objective function and its gradient are more complex than in Eq. (17.4). For example, the gradient, $\nabla_i S$, requires the calculations of the second derivatives of the potential. The overall computational cost of evaluating the right-hand side of Eq. (17.8) is higher than the computational cost of the right-hand side of Eq. (17.4).

From the discussion so far, we have learned how to compute a set of points that approximate a minimum energy or a free energy pathway. The curve connects two structures (or states in the case of free energy). One structure represents the reactant and the other represents the product (Figure 17.1). It is, however, still unclear what exactly are we going to do with the curve once it is determined? The prime goal of this monograph is, after all, to study time scales and kinetics.

There are a number of ways to exploit reaction paths in rate calculations. The most obvious use is the identification of bottlenecks. A reaction path charts a way to reach the product from the reactant, which is the least costly from the perspective of energy or free energy. The highest energy (free energy) point along the path is the barrier the system needs to cross to reach a desired state. If the barrier is much higher than any other barriers found along the pathway, then statistical approaches to the calculations of kinetics, such as Transition State Theory (TST [7], see also Chapter 8) are appropriate. In TST the focus is on a small domain in space, which is the bottleneck. It therefore does not exploit the full information generated in the reaction path calculations.

The identification of the highest energy point along a minimum energy path is useful if this energy barrier dominates the kinetics. This maximum along the minimum energy path is a saddle point in which the energy is minimized in all directions with the exception of the direction of the path. The determination of a complete pathway seems like an overkill if the intention is to identify just one point. Indeed, there are computational approaches that focus on the calculations of a saddle point to be used in TST calculations rather than on the calculations of the entire path.

Are there benefits to a complete pathway versus a transition state for the purpose of rate determination? If the system has one dominant barrier, then there are none. However, in many condensed phase simulations the energy or free energy landscapes are rough and include a large number of minima and barriers. It is not always possible to determine a single dominant barrier and different approaches are required.

If there is a distribution of saddle points along the path, with numerous barriers of similar heights (Figure 16.4), more than a single transition state is required, and it is not at all clear that TST analysis is helpful. TST is based on the concept of "no return" once a barrier is crossed. If returns are likely as is the case with multiple comparable barriers, crossing and re-crossing are likely to be found, and the Transition State Theory is inaccurate. While it is possible to correct the transition state theory with the transmission coefficient (Chapter 8.2.4 and reference [8]), the corrections require additional calculations that may be costly, especially if multiple bottlenecks are present.

The curves computed according to Eq. (17.4) or (17.8) chart a path of least resistance between the two states. It is therefore likely that thermal trajectories will follow roughly through the pre-determined route. Compare for example the path in Figure 17.1 and the trajectory in Figure 16.2. From a computational perspective, the reaction path is a way to narrow the search for plausible reactive trajectories, making it possible to enhance the sampling of productive events. The set of discrete points that makes the minimum energy path helps the calculations of trajectories in the neighborhood of the reaction path as explained in the next section (Celling). Guiding the sampled space to the neighborhood of the path helps build statistics of reactive trajectories more efficiently. Once statistics of reactive trajectories are computed, the MFPT and other kinetic observables can be calculated as well. This is the topic of the next Chapter.

We comment that focusing on the selection of reactive trajectories is at the core of the Transition Path Sampling method (TPS). [9] In principle, the TPS avoids the use of a reaction coordinate and samples instead reactive trajectories between reactant and product. It relies, however, on the availability of a guiding reactive trajectory accessible to conventional MD (i.e. a short trajectory) that seeds the sampling. Such a guiding trajectory can be a reaction coordinate. Indeed, other approaches that were developed based on the TPS, such as the Transition Interface Sampling (TIS [10]), employ a reaction coordinate or an order parameter to guide the trajectories to desired outcomes.

References

1 Ciccotti, G., Kapral, R., and Vanden-Eijnden, E. (2005). Blue moon sampling, vectorial reaction coordinates, and unbiased constrained dynamics. *ChemPhysChem* 6 (9): 1809–1814.

2 Templeton, C., Chen, S.H., Fathizadeh, A. et al. (2017). Rock climbing: a local-global algorithm to compute minimum energy and minimum free energy pathways. *J. Chem. Phys.* 147 (15): 10.

3 Ulitsky, A. and Elber, R. (1990). A new technique to calculate steepest descent paths in flexible polyatomic systems. *J. Chem. Phys.* 92 (2): 1510–1511.

4 Elber, R. and Karplus, M. (1987). A method for determining reaction paths in large molecules – application to myoglobin. *Chem. Phys. Lett.* 139 (5): 357–380.

5 Weinan, E., Ren, W.Q., and Vanden-Eijnden, E. (2002). String method for the study of rare events. *Phys. Rev. B* 66 (5): 4.

6 Jonsson, H., Mills, G., and Jacobson, K.W. (1997). Nudged elastic band method for finding minimum energy paths of transitions. In: *Classical and Quantum Dynamics in Condensed Phase Simulations* (ed. B.J. Berne, G. Ciccotti, and D.F. Coker). Singapore: World Scientific.

7 Vanden-Eijnden, E. and Tal, F.A. (2005). Transition state theory: variational formulation, dynamical corrections, and error estimates. *J. Chem. Phys.* 123 (18).

8 Chandler, D. (1978). Statistical-mechanics of isomerization dynamics in liquids and transition-state approximation. *J. Chem. Phys.* 68 (6): 2959–2970.

9 Dellago, C., Bolhuis, P.G., and Geissler, P.L. (2005). Transition path sampling. In: *Advances in Chemical Physics*, vol. 123, 1–78. New York: Wiley.

10 van Erp, T.S. and Bolhuis, P.G. (2005). Elaborating transition interface sampling methods. *J. Comput. Phys.* 205 (1): 157–181.

18

Celling

A modern idea for computing long-time molecular kinetics with trajectories is the use of cells in coarse space. Instead of following the process of interest as the progression of points in phase space (conventional trajectories) we partition the phase space into cells and focus on transitions between cells. We compute descriptors of local kinetics between the cells and develop a model to obtain the overall kinetics and thermodynamics. We call the process of space partition and kinetic modeling using cells – "celling". An example for a specific partitioning to cells is shown in Figure 18.1. This approach may superficially look like the ME for MS-s. Note, however, that we did not require local equilibrium at the different cells, which make "celling", in principle, an exact procedure and different from the ME.

The idea of "celling" was used in the past, even though not in the computationally intensive way it is conducted today. For example, one interpretation of the TST is the partition of phase space into two cells, one cell of reactants and one cell of products. The transition state is at the boundary of the cells. The boundaries play a critical role in the theory of cell kinetics. Furthermore, when kinetic modeling invokes a number of intermediates between the reactant and product, we may think on the intermediates as another realization of cells. In the last case, metastable states can be made into cells. However, the reverse is not true, cells do not have to be metastable states.

The application of celling requires two critical steps. The first is to define the cell geometrically, and the second is to determine a measure for transitions between the cells. There are several theories and algorithms that share the concept of celling but define cells and the transitions between them differently. Approaches to define cells include clusters of MD trajectories [1], eigenvectors of a discrete kinetic matrix [2], and modeling of continuous diffusive process [3]. These definitions rely on prior trajectory data that is analyzed to determine the state geometries. For example, clustering the trajectory shown in Figure 16.2 suggests the existence of two cells: a cell for each metastable state.

However, the requirement of calculations of trajectories prior to the analyses is expensive. Some of the transitions between cells are rare events and

Molecular Kinetics in Condensed Phases: Theory, Simulation, and Analysis,
First Edition. Ron Elber, Dmitrii E. Makarov and Henri Orland.
© 2020 John Wiley & Sons Ltd. Published 2020 by John Wiley & Sons Ltd.

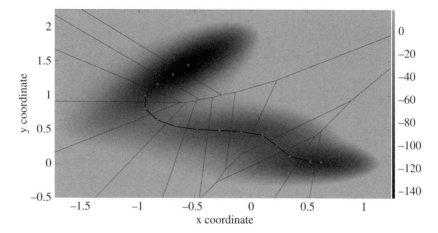

Figure 18.1 The Mueller potential and its partition to Voronoi cells using configurations along a reaction coordinate. The black line is a discrete reaction coordinate computed with Eq. (17.3). The blue dots along the line are the discrete optimized points that are also used as centers of Voronoi cells. The green straight-line segments are the boundaries of the Voronoi cells that are also used as milestones. (*See color plate section for color representation of this figure*).

challenging to sample, posing a significant computational burden. A key advantage of the reaction path approach is the ability to generate plausible partitions of the reaction space without the considerable investment in computing first long-time trajectories and regardless of the rarity of the transition (another technique to avoid the calculations of trajectories is the Gaussian modeling of phase space densities [4]).

One approach to map a reaction path to cells utilizes hyperplanes orthogonal to it. We construct cells along the reaction coordinate by inserting hyperplanes orthogonal to the reaction coordinate at specific points. Let Φ_i be a point along the reaction coordinate and let $\mathbf{e}_i$ be a unit vector along the RC at Φ_i. The set of points $\mathbf{y}$ such that $(\mathbf{y} - \Phi_i)^t \cdot \mathbf{e}_i = 0$ form the desired hyperplane. The space bound by two hyperplanes (in the one dimension of the RC) is considered a cell. The hyperplane model is intuitive and was used extensively in the context of TST. However, it is valid only in the neighborhood of the line or the curve that represents the RC. Far from the RC curve different hyperplanes may cross, making the assignment of conformations to cells ambiguous.

Vanden Eijnden suggested an alternative use of RC configurations to identify cells. The set of discrete points along the reaction coordinate are made into centers of Voronoi cells. [5] A Voronoi cell is defined with a distance criterion. A point $\mathbf{y}$ is in a Voronoi cell i if the distance $|\mathbf{y} - \Phi_i|$ between the point $\mathbf{y}$ and the center of cell, Φ_i, is smaller than the distances $|\mathbf{y} - \Phi_j|\ j \neq i$ to all other cell centers - Φ_j. The mapping of configurations or phase space points to Voronoi cells is unique. Every phase space point is assigned to a cell or to an interface between cells without ambiguity. This assignment is therefore better than the use of hyperplanes orthogonal to the RC that may cross. Voronoi tessellation is valid for any reaction coordinate and for an arbitrary distance from the discrete set of configurations along the pathway. It is actually valid for any set of points that are used as centers of Voronoi cells to represent the reaction space, and is not limited to RC. The use of the RC, however, helps placing the cells where they matter the most.

The classification to cells by Voronoi tessellation of RC structures is fundamentally different from a classification using clustering of trajectories. The mapping of the RC configurations is static and geometrical. No information on the system dynamics and metastability is built into the construction. This makes the Voronoi cell approach less effective than trajectory analysis in detecting metastability. On the other hand, it makes it easier to sample rare events by placing (for example) a cell on the top of a free energy barrier. Obviously, barrier tops are not metastable states and transitions between the cell at the top of the barrier and other cells are rapid and may be faster than relaxation times within the cell. There is no reason to expect that modeling dynamics with this type of cells with the Master Equation (assuming a Markov process) will be successful. We need different types of theories and algorithms to investigate the dynamics of cells that are not in local equilibrium and do not offer separation of time scales. If we are able to formulate such an algorithm the advantages will be substantial. We will be able to use small (unstable) cells at the tops of energy barriers which will accelerate the computations of rare events. As we argue below, the more cells we have the more parallel is the algorithm. Hence, it will run more efficiently on modern (parallel) computer architecture.

Indeed, there is a class of different theories and algorithms that consider the *flux* between cells instead of cell probability as the core function to study kinetics. [6–9] The flux is defined as the number of trajectories that passes a boundary between cells per unit time. We denote the flux by q and we seek an equation that determines the flux based on trajectory information. The discussion below follows the theory of Milestoning. [10]

Consider a trajectory, $x(t)$, that passes exactly at time τ a boundary M_α between two Voronoi cells. We call the boundary between the cells a milestone. Here we use x and y to denote phase space points. In Fig 18.1 a green straight line presents a milestone. The last milestone a trajectory crosses defines the coarse state of the system. For example, if the last milestone crossed is M_α we say that the trajectory is in a state α. Spatially, the coarsening definition means that as long as the trajectory is in state α it can be in any cell that includes milestone α as a boundary. To derive an equation for the fluxes at the milestones, we ask, "What was the state of the trajectory before it crosses milestone α at exactly time τ?"

$$q_\alpha(\mathbf{x}, \tau) = \delta(\tau)p_0(\alpha, \mathbf{x}) + \sum_\beta \int_0^\tau \int_{M_\beta} dt \cdot d\mathbf{y} \cdot q_\beta(\mathbf{y}, t) \cdot w_{\beta,\alpha}(\mathbf{y}, t; \mathbf{x}, \tau)$$

(18.1)

Eq. (18.1) is a simple counting of trajectories. We count the number of trajectories that pass a particular milestone and trace their history. This gives us a consistent equation for the fluxes. Eq. (18.1) has the following verbal interpretation. On the left hand side of the equation we count the number of trajectories that pass a phase space point $\mathbf{x}$ in milestone α at time τ per unit time ($q_\alpha(\mathbf{x}, \tau)$). The r.h.s. of the equation sums up all the channels that are leading to the trajectory entry to state α. The trajectory may just happen to be at milestone α at time zero according to the initial conditions (the term $\delta(\tau)p_0(\alpha, \mathbf{x})$). Alternatively, the trajectory that crosses milestone α may have come from other nearby milestones that are indexed by β_i (Figure 18.2, lower panel). The trajectories from β_i do not cross any other milestone β' before crossing milestone α. However, they are allowed to re-cross the milestone β_i on which they were initiated. The trajectory β_3 which starts at lower panel in Figure 18.2 illustrates a re-crossing of the initiating milestone.

The fluxes of trajectories at each of the milestone β of Eq. (18.1) are given by $q_\beta(\mathbf{y}, t)$. Given that a trajectory was initiated at β at time t and a phase space point $\mathbf{y}$, the probability that it will make it to milestone α, phase space point $\mathbf{x}$, at time τ is the kernel $w_{\beta\alpha}(\mathbf{y}, t; \mathbf{x}, \tau)$. The final summation and integrals over all milestones, phase space points, and earlier times are to ensure that we include all possible trajectories that made it to α at time τ.

The kernel above is similar to the one we were considering during the ME discussion (Chapter 14). For the ME, the kernel provides approximate dynamics since the propagated states are not PS-s. In contrast, Eq. (18.1) is exact and is valid for numerous models of the dynamics, provided that a transition kernel can be defined and computed. We estimate elements of the kernel matrix by trajectories between milestones.

The unknown and the prime function that we are interested in is the flux $q_\alpha(\mathbf{x}, \tau)$. Eq. (18.1) is a linear equation for fluxes that uses as inputs the initial conditions (the distributions at time zero) and the kernel. So, in principle, Eq. (18.1) can be solved with standard linear solvers. However, the computational efforts to find an exact solution can be prohibitively expensive. In particular the cost to obtain an explicit expression for the exact kernel is enormous. We need to determine the transition probability to every phase space point in any milestone that can be accessed directly from any phase point in the initiating milestone. There are several approaches to reduce the complexity of this general approach and still obtain useful kinetic and thermodynamic observables, which we discuss below.

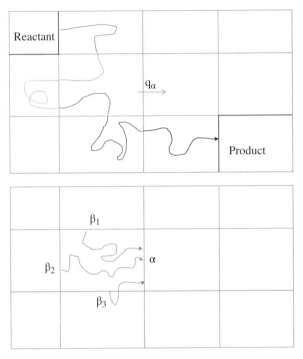

Figure 18.2 Top: A schematic representation of Milestoning. The space is partitioned to cells and trajectories are conducted between boundaries of cells. The quantity of interest is the flux, q_α, which is the number of trajectories that cross milestone α in unit time. It is also possible to consider a single long reactive trajectory and to chop it into pieces between boundaries as illustrated for the colored trajectory fragments. Bottom: A schematic representation of Milestoning trajectories that are used to estimate the flux in Eq. (18.1). Trajectories are initiated at milestones β_1, β_2, β_3 and their time courses are followed until they hit for the first time milestone α. Note that the trajectories are allowed to cross the milestone they were initiated at. The total number of trajectories that cross milestone α at time τ is the summation over all the trajectories that arrived to it for the first time at τ. Using the trajectories, we can estimate the kernel $w_{\beta, \alpha}(\mathbf{y}, t; \mathbf{x}, \tau)$. (*See color plate section for color representation of this figure*).

First, we assume that the transition probability depends only on the time difference, $w_{\beta, \alpha}(\mathbf{y}, t; \mathbf{x}, \tau) = w_{\beta, \alpha}(\mathbf{y}, \mathbf{x}; \tau - t)$, that is, the process is homogeneous in time. This is frequently the case in molecular systems if the system does not experience an external time-dependent force.

Second, we consider the stationary solution of Eq. (18.1). This is similar in spirit to the steady state conditions of a chemical reaction in which the change in *concentration* (or flux) of an intermediate is approximately zero. [11] A stationary solution can be obtained by the long-time limit of a system that includes a source of reactants (reactants are provided to the system at a constant rate) and a sink (products are removed from the system) that balances the source. The system retains non-zero flux at all times. At the long time limit the flux becomes time independent, $\lim_{t \to \infty} q_\alpha(t) = q_{\alpha,stat}$ (We do not consider sustained non-equilibrium oscillations that can be caused, for example, by an external and time-dependent field). We have

$$\lim_{\tau \to \infty} \left[q_\alpha(\mathbf{x}, \tau) = \delta(\tau) p_0(\alpha, \mathbf{x}) + \sum_\beta \int_0^\tau \int_{M_\beta} dt' \cdot d\mathbf{y} \cdot q_\beta(\mathbf{y}, t') \right.$$

$$\left. \cdot w_{\beta,\alpha}(\mathbf{y}, \mathbf{x}; \tau - t') \right] =$$

$$q_{\alpha,stat}(\mathbf{x}) = \sum_\beta \int_{M_\beta} d\mathbf{y} \cdot q_{\beta,stat}(\mathbf{y}) \cdot \int_0^\infty dt' \cdot w_{\beta,\alpha}(\mathbf{y}, \mathbf{x}; t')$$

$$q_{\alpha,stat}(\mathbf{x}) = \sum_\beta \int_{M_\beta} d\mathbf{y} \cdot q_{\beta,stat}(\mathbf{y}) \cdot w_{\beta,\alpha}(\mathbf{y}, \mathbf{x}) \tag{18.2}$$

We are using the property that $w_{\beta,\alpha}(\mathbf{y}, \mathbf{x}; t)$ decays rapidly as a function of the time, t and therefore most of the contribution to the time integral is when the flux is already stationary. We define in the last line a time-independent kernel similar to Eq. (14.4). Alternatively, Eq. (18.2) can be derived with the Laplace transform. [10] Note that our current expression is exact and there is no need to assume metastability in the α states like we have done when discussing the Master Equation. The milestones are used to partition the space and label the trajectories, but not necessarily to approximate the dynamics.

Eq. (18.2) is a linear equation from which we eliminate the time dependence of Eq. (18.1). Formally it can be written as a vector-matrix equation $\mathbf{q}^t = \mathbf{q}^t\mathbf{w}$ in a very high dimension. In this compact vector matrix notation $q_\alpha(\mathbf{x})$ is a long one-dimensional vector with two indices: the milestone index, and the position in the milestone plane.

Let us assume for the moment that we were able to solve the linear equation and we have $q_\alpha(\mathbf{x})$ at hand. How can we continue to determine the thermodynamics and kinetics of the system?

The probability of being at state α at time τ is the probability of entering that state at an earlier time t and not leaving it until time τ, we write

$$P_\alpha(\mathbf{x}, \tau) = \int_0^\tau dt \cdot q_\alpha(\mathbf{x}, t) \left[1 - \sum_\beta \int_{M_\beta} d\mathbf{y} \int_0^{\tau - t} dt' \cdot w_{\alpha\beta}(\mathbf{x}, \mathbf{y}; t') \right] \tag{18.3}$$

Eq. (18.3) has the following verbal interpretation. The probability that the last crossing point at times earlier than τ was at coordinate $\mathbf{x}$ of milestone α (the point $(\alpha, \mathbf{x}, \tau)$ in time and space) is the left hand side expression $(P_\alpha(\mathbf{x}, \tau))$. A detailed calculation of this probability is provided at the right side. To be at

$(\alpha, \mathbf{x}, \tau)$ we must cross α at position $\mathbf{x}$ at an earlier time t. This flux of crossing trajectories at earlier times is given by $q_\alpha(\mathbf{x}, t)$ $t < \tau$. After the entry event and before time τ is reached, some of the trajectories may pass another milestone and leave the state of interest. These trajectories should not be counted as in the state $(\alpha, \mathbf{x}, \tau)$. The matrix element $w_{\alpha\beta}(\mathbf{x}, \mathbf{y}; t')$ counts exits from the state $(\alpha, \mathbf{x})$ to state $(\beta, \mathbf{y})$ at time t' between 0 and $\tau - t$. Summing up over all exit states $(\beta, \mathbf{y})$ and times t' between 0 and $\tau - t$ we obtain the probability lost to other states $\sum_\beta \int_{M_\beta} d\mathbf{y} \int_0^{\tau-t} dt' \cdot w_{\alpha\beta}(\mathbf{x}, \mathbf{y}; t')$. The probability to remain at

$(\alpha, \mathbf{x})$ is $\left[1 - \sum_\beta \int_{M_\beta} d\mathbf{y} \int_0^{\tau-t} dt' \cdot w_{\alpha\beta}(\mathbf{x}, \mathbf{y}; t') \right]$. Multiplying the flux of trajectories entering to $(\alpha, \mathbf{x})$ at earlier times by the probability that they will remain in the same state and summing over all earlier times t we obtain the right-hand side expression of Eq. (18.3).

The stationary or the long-time solution to Eq. (18.3) is obtained by computing the Laplace transform (see Eq. (14.16)) of Eq. (18.3) and considering the limit of a small Laplace variable, which is equivalent to long time. The final result is derived in Eq. (18.4) below. It is the stationary flux of entering state α multiplied by the lifetime $t_\alpha(\mathbf{x})$ of milestone α. The lifetime of a milestone α is the average time (averaged over multiple trajectories) it takes a trajectory initiated at milestone α at phase space point $\mathbf{x}$ to hit for the first time a milestone different from α. Hence, to obtain the probability it is not enough to know the stationary flux, we are also required to know the lifetime of the milestone.

$$\lim_{\lambda \to 0} \left\{ \tilde{P}_\alpha(\mathbf{x}, \lambda) = \tilde{q}_\alpha(\mathbf{x}, \lambda) \left[\frac{1}{\lambda} \left(1 - \sum_\beta \int_{M_\beta} d\mathbf{y} \cdot \tilde{w}_{\alpha\beta}(\mathbf{x}, \mathbf{y}; \lambda) \right) \right] \right\}$$

$$P_{\alpha,stat}(\mathbf{x}) = q_{\alpha,stat}(\mathbf{x}) \cdot \lim_{\lambda \to 0} \left\{ \frac{1}{\lambda} \left[1 - \sum_\beta \int_{M_\beta} d\mathbf{y} \cdot \int_0^\infty \exp(-\lambda t) w_{\alpha\beta}(\mathbf{x}, \mathbf{y}; t) dt \right] \right\}$$

$$= q_{\alpha,stat}(\mathbf{x}) \cdot \lim_{\lambda \to 0} \left\{ \frac{1}{\lambda} \left[1 - \sum_\beta \int_{M_\beta} d\mathbf{y} \cdot \int_0^\infty dt \cdot w_{\alpha\beta}(\mathbf{x}, \mathbf{y}; t) \right. \right.$$

$$\left. \left. + \sum_\beta \int_{M_\beta} d\mathbf{y} \int_0^\infty dt \cdot (\lambda t) w_{\alpha\beta}(\mathbf{x}, \mathbf{y}; t) \right] \right\}$$

$$P_{\alpha,stat}(\mathbf{x}) = q_{\alpha,stat}(\mathbf{x}) \sum_\beta \int_{M_\beta} d\mathbf{y} \cdot \int_0^\infty t \cdot w_{\alpha\beta}(\mathbf{x}, \mathbf{y}; t) \cdot dt$$

$$P_{\alpha,stat}(\mathbf{x}) = q_{\alpha,stat}(\mathbf{x}) t_\alpha(\mathbf{x}) \tag{18.4}$$

The free energy of state α, $F_\alpha(\mathbf{x})$, is given by $F_\alpha(\mathbf{x}) = -k_B T \log[P_\alpha(\mathbf{x})]$. The system is not in equilibrium since it includes a source and a sink. Therefore, the physical meaning of the "free energy" here is not obvious. However, it can be used if the overall flux is small and the system is in an "almost" (or local) equilibrium state.

The last entity that we wish to derive is the mean first passage time (MFPT), but wait… we already derived an expression for the MFPT in Eq. (14.11). We rewrite it here in a form more consistent with the Milestoning formulation

$$\langle \tau_{\alpha\beta}(\mathbf{x},\mathbf{y})\rangle = t_\alpha(\mathbf{x}) + \sum_\gamma \int_{M_\gamma} d\mathbf{z}\cdot w_{\alpha\gamma}(\mathbf{x},\mathbf{z})\langle \tau_{\gamma\beta}(\mathbf{z},\mathbf{y})\rangle \tag{18.5}$$

Eq. (18.5) is a linear equation for the overall MFPT initiating trajectories at α and terminating them at another milestone β. Equations (18.1–18.5) summarize the exact formulation of Milestoning. Given the kernel, we are able to determine functions of interest, the probability of the state, the free energy of a state and the mean first passage time between states of choice. An application is described in Chapter 19 and several concrete studies were reviewed recently. [12]

We consider in more details the calculations of the flux. One approach to compute the integrals found on the r.h.s. of Eq. (18.2) is by Monte Carlo sampling. We write without loss of generality $q_\alpha = \theta_\alpha f(\mathbf{x}_\alpha)$ where θ_α is the milestone weight, $\mathbf{x}_\alpha$ is the phase space point in the milestone, and $f(\mathbf{x}_\alpha)$ is a first hitting point probability density function which is normalized to one: $\int_{M_\alpha} f(\mathbf{x}_\alpha)\cdot d\mathbf{x}_\alpha = 1$. It is the probability that a trajectory that hits milestone α for the first time, will hit it at phase space point $\mathbf{x}_\alpha$.

We integrate over $\mathbf{x}_\alpha$ on both sides of the lowest line of Eq. (18.2) to have

$$\int_{M_\alpha}\theta_\alpha f_\alpha(\mathbf{x}_\alpha)d\mathbf{x}_\alpha = \sum_\beta \int_{M_\alpha}\int_{M_\beta} d\mathbf{x}_\alpha d\mathbf{x}_\beta \theta_\beta f_\beta(\mathbf{x}_\beta)w_{\beta\alpha}(\mathbf{x}_\beta,\mathbf{x}_\alpha)$$

$$\theta_\alpha = \sum_\beta \theta_\beta \int_{M_\alpha}\int_{M_\beta} d\mathbf{x}_\alpha d\mathbf{x}_\beta f_\beta(\mathbf{x}_\beta)w_{\beta\alpha}(\mathbf{x}_\beta,\mathbf{x}_\alpha)$$

$$\theta_\alpha = \sum_\beta \theta_\beta \overline{w}_{\beta\alpha}$$

$$\overline{w}_{\beta\alpha} = \int_{M_\alpha}\int_{M_\beta} d\mathbf{x}_\alpha d\mathbf{x}_\beta f_\beta(\mathbf{x}_\beta)w_{\beta\alpha}(\mathbf{x}_\beta,\mathbf{x}_\alpha) \tag{18.6}$$

The formula in the third line of Eq. (18.6) is a linear equation for the unknown Milestoning weights, θ. The number of milestones is much lower than the system dimensionality which makes the calculations of the $\theta-s$ straightforward. Typically, we have only a few hundred to a thousand milestones compared to hundreds of thousands of particles in an atomically detailed simulation. The average kernel is computed as a sum over all termination points in milestone α and as a weighted average (with probability density $f_\beta(\mathbf{x}_\beta)$) over initiating points.

To begin with, the $f_\beta(\mathbf{x}_\beta)$ are not known and we solve for them by iterations. For the first iteration we set $f_\beta^{(1)}(\mathbf{x}_\beta) = \frac{\exp[-\beta H(\mathbf{x}_\beta)]}{Z'_\beta}$, $Z'_\beta = \int_{M_\beta} d\mathbf{x}_\beta \exp[-\beta H(\mathbf{x}_\beta)]$ where Z'_β is the normalization in the milestone hypersurface. The functional form of the first iteration is equivalent to assuming a local thermal equilibrium in the milestone. This assumption is similar to what we invoked for metastability of states. Deviations from metastability inside the milestones can be captured later by additional iterations (Eq. (18.9)). For now, we integrate over $\mathbf{x}_\alpha$ on both sides of Eq. (18.6) with the explicit expression for $f_\beta^{(1)}(\mathbf{x}_\beta)$ to obtain

$$\int_{M_\alpha} d\mathbf{x}_\alpha \theta_\alpha^{(1)} \frac{\exp[-\beta H(\mathbf{x}_\alpha)]}{Z'_\alpha} =$$

$$= \sum_\beta \int_{M_\alpha} \int_{M_\beta} d\mathbf{x}_\alpha d\mathbf{x}_\beta \theta_\beta^{(1)} \frac{\exp[-\beta H(\mathbf{x}_\beta)]}{Z'_\beta} w_{\beta\alpha}(\mathbf{x}_\beta, \mathbf{x}_\alpha)$$

$$\theta_\alpha^{(1)} = \sum_\beta \theta_\beta^{(1)} \overline{w}_{\beta\alpha}^{(1)} \tag{18.7}$$

where the averaged kernel is

$$\overline{w}_{\beta\alpha}^{(1)} = \int_{M_\beta} \int_{M_\alpha} d\mathbf{x}_\alpha d\mathbf{x}_\beta \frac{\exp[-\beta H(\mathbf{x}_\beta)]}{Z'_\beta} w_{\beta\alpha}(\mathbf{x}_\beta, \mathbf{x}_\alpha) \tag{18.8}$$

The kernel of Eq. (18.8) is an average over trajectories that start at milestone β with a canonical weight and end anywhere at milestone α. An element of the averaged kernel is estimated by trajectory sampling. The initial coordinates and velocities of the trajectories are sampled according to the weight $\exp[-\beta H(\mathbf{x}_\beta)]$. The trajectories are integrated in time until they hit for the first time another milestone. Let the number of initial trajectories at milestone β be n_β. Let the number of trajectories initiated at β and terminated at α (hit the α milestone for the first time) be $n_{\beta\alpha}$. The averaged kernel is estimated as $\overline{w}_{\beta\alpha} = n_{\beta\alpha}/n_\beta$. With the averaged kernel at hand and with the help of Eq. (18.7) we determine the milestone weights, θ_α and the explicit functional form for the flux vector $\mathbf{q}$.

The trajectories provide more information than the averaged kernel. We also obtain new hitting points at $\mathbf{x}_\alpha$ and an estimate of $f_\alpha(\mathbf{x}_\alpha)$ from all the Milestoning trajectories that arrive at α from nearby milestones. The hitting points that we collect at the milestones may be distributed differently from the canonical (equilibrium) distribution we assumed to begin with. We write

$$f_\alpha^{(2)}(\mathbf{x}_\alpha) = \frac{1}{\theta_\alpha^{(1)}} \sum_\beta \int_{M_\beta} d\mathbf{x}_\beta \theta_\beta^{(1)} f_\beta^{(1)}(\mathbf{x}_\beta) w_{\beta\alpha}(\mathbf{x}_\beta, \mathbf{x}_\alpha) \tag{18.9}$$

The state of the trajectory is determined by the last milestone it crosses, not by the instantaneous coordinate vector. Therefore, the distribution of termination points at the milestone may not be canonical even if the system is in equilibrium. The new, numerically generated distribution function of first hitting points can be used in another iteration of the Milestoning calculations replacing $f^{(1)}$ by termination points collected at the milestone, which we call $f^{(2)}$, and recalculating the milestones' weights according to Eq. (18.6). The iterations repeat using the new $f^{(n)}$ until convergence is reached. Convergence is rigorously defined when the difference between $f^{(n+1)}$ and $f^{(n)}$ is smaller than an allowed error value. We showed that the process is guaranteed to converge if the MFPT is finite due to favorable properties of the kernel. [10, 13] The rate of convergence is determined by the difference between the largest eigenvalue of **w**, which is one, and the next largest eigenvalue with a norm smaller than one. There is a complication in our ability to obtain sufficient statistics at all milestones when considering subsequent iterations. It is that the terminating points are not necessarily distributed uniformly between the milestones, in contrast to the first iteration. A possible remedy is to run more trajectories at "offending" milestones. Since the trajectories are short running more trajectories at bottlenecks should be feasible.

The distribution functions in the milestone may be difficult to assess since they can be of high dimension. They are functions of all atomic degrees of freedom minus the degrees of freedom in coarse space that are used to define the milestones. Therefore, we typically probe the convergence of critical averaged observables such as the MFPT or the free energy as a function of the iteration number to decide when to stop the iterations rather than probing distributions in high dimension.

In the past, a single iteration of the probability density, f, was used to obtain approximate results [14]. The approximation of one iteration is valid when the milestones are reasonably well separated from each other allowing for sufficient time to reach a local equilibrium at the milestone. The significant simplification obtained in this case makes it possible for us to write the free energy and the mean first passage time as a function of the milestone index only with a single set of trajectories. This is similar in spirit to the local equilibrium assumption of the Master equation (Eqs. (8.8) and (14.1)). However, Milestoning makes it possible to probe non-equilibrium flux in the coarse space and to refine the results (by additional iterations) to the exact results. We are not aware of iterative procedure that makes the Master Equation exact.

Sampling conformations to initiate trajectories from the canonical distribution and using a single iteration we have for the free energy

$$F_\alpha = -k_B T \log[P_\alpha] \tag{18.10}$$

and the MFPT:

$$\langle \tau \rangle = \mathbf{p}(0)(1 - \mathbf{w})^{-1}\mathbf{t} \tag{18.11}$$

Where $\mathbf{p}(0)$ is the initial state vector, Eq. (18.11) provides a vector of MFPT-s from any milestone to the absorbing milestone.

The estimate of the statistical errors in the calculation of the kernel with trajectories is important, and below we describe the approach we use. Another way of thinking about the matrix element $\overline{w}_{\beta\alpha}$ is as an average of outcomes of trajectories, which can be zeroes (did not reach milestone α) or ones (reached milestone α). Hence, in this picture the kernel matrix element is

$$\overline{w}_{\beta\alpha} = \frac{1}{n_\beta} \sum_{l=1}^{n_\beta} \delta_{\gamma(l)\alpha} \tag{18.12}$$

where $\delta_{\gamma(l)\alpha}$ is the Kronecker delta function. It is zero if the terminating milestone $\gamma(l)$ of a trajectory l is different from α and it is one if $\gamma(l) = \alpha$. Hence, the entries to the sum are random numbers, assumed independent, which are either zero or one. Therefore the kernel element, $\overline{w}_{\beta\alpha}$, is also a random number which is sampled from the β distribution [15, 16].

$$P(\overline{w}_{\beta\alpha}) = \frac{\Gamma(n_\beta)}{\Gamma(n_\beta - n_{\beta\alpha})\Gamma(n_{\beta\alpha})} \overline{w}_{\beta\alpha}^{\,n_{\beta\alpha}-1} (1 - \overline{w}_{\beta\alpha})^{n_\beta - n_{\beta\alpha}-1} \tag{18.13}$$

where $\Gamma(x)$ is the gamma function. To estimate the errors in the flux, free energy, and MFPT we repeat the calculations for an ensemble of kernels, $\overline{w}_{\beta\alpha}$, sampled according to Eq. (18.13). The parameters for the distribution, n_β and $n_{\beta\alpha}$, (the number of trajectories that were initiated at milestone β, and the number of trajectories initiated at β and terminated at α, respectively) are estimated from the trajectories. Each value of observable is computed from one sample of the kernel. The ensemble of kernels therefore provides a distribution of observable values from which the error bars are computed.

We summarize below our discussions on the use of trajectories in the calculations of kinetic observables. In the most straightforward use of trajectories in rate calculations, we initiate them at the reactant state and integrate the equations of motion (e.g. Eq. (15.1)) until they enter the product state for the first time. Then, we average the arrival times to the product state of the entire ensemble of trajectories to obtain the MFPT (Eq. (13.6)). We can also extract the rate coefficient by examining the decay of the probability of the initial state and plotting $\log(P)$ versus the time (Figure 16.6).

The second perspective, which is appropriate for a system in equilibrium, uses a single trajectory that moves forwards and backwards between the reactant and product states (Figures 16.2 and 16.3). We chop this trajectory to segments. Each of the segments includes only one transition between the reactants and products and use these trajectory segments to estimate (again) the MFPT and the rate.

The power of these reactive trajectories is that they are general in their applicability and can be used to model many types of events that can be non-Markovian and/or non-equilibrium processes. However, they can be prohibitively expensive if the reaction is slow and/or the transition is a rare event.

For rare events and slow processes, we propose that the use of short trajectories between boundaries of cells in coarse space instead of computing complete reactive trajectories. The Milestoning approach can lead to tremendous saving in computational resources since we can place the cells at the tops of the barriers gaining significant statistics of rare transitions. The cells are distributed to cover the relevant reactive coarse space. The calculations of transitions between the cells allow modeling of the entire kinetics.

The "cell"-based computations can be conducted in hierarchical fashion, starting from an approximate local equilibrium estimate for the distribution in the milestone, and expanding the scope of the calculations to include exact first hitting point distributions (*FHD*). Estimating the *FHD* is more expensive than to conduct local equilibrium studies. This is since FHD requires trajectory history while equilibrium sampling only needs configurations. However, use of cells is still more efficient than conventional and complete trajectory calculations from reactants to products. We believe that the combination of ensembles of short trajectories and kinetic modeling of transition between cells will continue to drive research into the kinetics of very complex processes [12].

References

1 Chodera, J.D., Singhal, N., Pande, V.S. et al. (2007). Automatic discovery of metastable states for the construction of Markov models of macromolecular conformational dynamics. *J. Chem. Phys.* 126 (15).

2 Sarich, M., Noe, F., and Schutte, C. (2010). On the Approximation Quality of Markov State Models. *Multiscale Model. Simul.* 8 (4): 1154–1177.

3 Boninsegna, L., Gobbo, G., Noe, F. et al. (2015). Investigating molecular kinetics by variationally optimized diffusion maps. HYPERLINK "https://pubs.acs.org/journal/jctcce" *J. Chem. Theory Comput.* 11 (12): 5947–5960.

4 Shalloway, D. (1996). Macrostates of classical stochastic systems. *J. Chem. Phys.* 105 (22): 9986–10007.

5 Vanden-Eijnden, E. and Venturoli, M. (2009). Markovian milestoning with Voronoi tessellations. *J. Chem. Phys.* 130 (19): 13.

6 Zhang, B.W., Jasnow, D., and Zuckerman, D.M. (2010). The "weighted ensemble" path sampling method is statistically exact for a broad class of stochastic processes and binning procedures. *J. Chem. Phys.* 132 (5).

7 Allen, R.J., Frenkel, D., and ten Wolde, P.R. (2006). Forward flux sampling-type schemes for simulating rare events: efficiency analysis. *J. Chem. Phys.* 124 (19): 17.

8 Moroni, D., Bolhuis, P.G., and van Erp, T.S. (2004). Rate constants for diffusive processes by partial path sampling. *J. Chem. Phys.* 120 (9): 4055–4065.

9 Dickson, A., Warmflash, A., and Dinner, A.R. (2009). Separating forward and backward pathways in nonequilibrium umbrella sampling. *J. Chem. Phys.* 131 (15).

10 Bello-Rivas, J.M. and Elber, R. (2015). Exact milestoning. *J. Chem. Phys.* 142 (9).

11 McQuarrie, D.A. and Simon, J.D. (1997). *Physical Chemistry: A molecular Approach*. Chapter 29.4. Sausalito, California:University Science Books.

12 Elber, R. (2017). A new paradigm for atomically detailed simulations of kinetics in biophysical systems. *Q. Rev. Biophys.* 50.

13 Aristoff, D., Bello-Rivas, J.M., and Elber, R. (2016). A mathematical framework for exact milestoning. *Multiscale Model. Simul.* 14 (1): 301–322.

14 West, A.M.A., Elber, R., and Shalloway, D. (2007). Extending molecular dynamics time scales with milestoning: example of complex kinetics in a solvated peptide. *J. Chem. Phys.* 126 (14).

15 Ma, P., Cardenas, A.E., Chaughari, M.L. et al. (2017). The impact of protonation on early translocation of anthrax lethal factor: kinetics from molecular dynamics simulations and milestoning theory. *J. Am. Chem. Soc.* 139: 14837–14840.

16 Mugnai, M.L. and Elber, R. (2015). Extracting the diffusion tensor from molecular dynamics simulation with milestoning. *J. Chem. Phys.* 142 (1): 18.

19

An Example of the Use of Cells: Alanine Dipeptide

We describe next a simulation of the conformational dynamics of alanine dipeptide in aqueous solution (Figure 19.1). Alanine dipeptide is a small system that is used to model local conformational transitions in the much larger protein backbone. Because it is small and relatively simple (even if embedded in aqueous solution with explicit representation of water molecules), it is also used to assess the performance of theoretical approaches to compute equilibrium and kinetics. Solvated alanine dipeptide can be investigated by a number of straightforward approaches (e.g. by a very long molecular dynamics trajectory, see Chapter 16) and it, therefore, makes it possible to compare accuracy and efficiency of different simulation techniques.

Conformational transitions in the backbone chain of peptides and proteins are frequently described by two torsion angles, φ and ψ, that are shown in Figure 19.1. These torsion angles are the coarse variables that we use in a Milestoning calculation. The free energy landscape of these two torsions (while integrating over all the solvent coordinates and the rest of the degrees of freedom of the peptide) is shown in Figure 19.2.

In figure 19.2 we draw 9 milestones as straight lines along the φ axis, keeping the value of ψ fixed. The nine milestones are equally spaced along ψ. This is the simplest choice and the positions of the milestones can be optimized for maximum computational efficiency. We conduct MD simulations restrained to a particular ψ value (a milestone) at each of the milestones to sample configurations for initial conditions from the distribution $f^{(1)}$ (Eq. 18.6). The simulations were conducted with the Molecular Dynamics package NAMD. [2] The water model was TIP3P. [3] We save a total of 120 coordinate sets at each milestone that were used to initiate unbiased trajectories between milestones. We integrate the equations of motion (Eq. (15.8)) with 1 femtosecond time step for a total of 100 picoseconds. The fixed length of time was easier to manage for multiple trajectories. The trajectories were then analyzed to determine Milestoning crossing events, the elements of the matrix **w** and the lifetimes **t**. We then use the transition probabilities and the milestone lifetimes to determine the free energy and the mean first passage time according to Eq. (18.10) and (18.11).

Molecular Kinetics in Condensed Phases: Theory, Simulation, and Analysis,
First Edition. Ron Elber, Dmitrii E. Makarov and Henri Orland.
© 2020 John Wiley & Sons Ltd. Published 2020 by John Wiley & Sons Ltd.

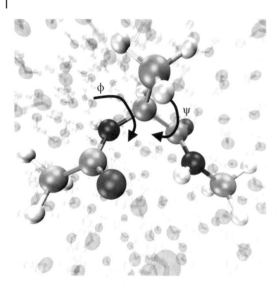

Figure 19.1 A space-filling model of an alanine dipeptide molecule embedded in a water box. The transparent pink spheres are the oxygen atoms of the water molecules. The big red and solid spheres are carbonyl oxygens and the blue spheres are nitrogen atoms. The rotation around bonds, φ and ψ are illustrated with the curved arrows and are the coarse variables that are typically used in studies of conformational transitions in protein and peptide backbones. The figure was prepared with the software VMD [1]. (*See color plate section for color representation of this figure*).

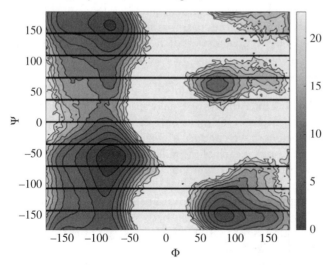

Figure 19.2 The free energy landscape of a solvated alanine dipeptide is shown as a function of the two coarse variables, φ and ψ. Note the existence of two deep minima on the left side of the map. The lower minimum on the left (negative ψ) corresponds to an α helix, a common secondary structure element of proteins, while the upper minimum on the left (positive ψ) is of an extended chain conformation (or a secondary structure of β sheet). The free energy depends only weakly on φ. Therefore, we reduce the number of coarse variables to one, ψ. Milestones along the ψ dihedral angle are shown as black lines parallel to the φ axis. (*See color plate section for color representation of this figure*).

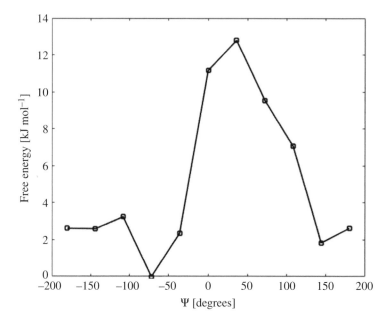

Figure 19.3 The free energy profile for a solvated alanine dipeptide conformational transition along the ψ dihedral angle computed with Milestoning.

The MFPT was 55.8 picoseconds suggesting rapid conversion between the two states, but still long enough to ensure local equilibrium.

We then determine the free energy profile, which is shown in Figure 19.3. It is consistent with Fig 19.2 in the sense that the transition from an α helix to a β sheet is more likely to happen by passing the periodic boundary (from -180 degrees to $+180$ degrees) than near 40 degrees. A similar calculation was conducted in [4].

References

1 Humphrey, W., Dalke, A., and Schulten, K. (1996). VMD: visual molecular dynamics, *J. Mol. Graph. Model.* 14 (1): 33–38.

2 Phillips, J.C., Braun, R., Wang, W. et al. (2005). Scalable molecular dynamics with NAMD. *J. Comput. Chem.* 26 (16): 1781–1802.

3 Jorgensen, W.L., Chandrasekhar, J., Madura, J.D. et al. (1983). Comparison of simple potential functions for simulating liquid water. *J. Chem. Phys.* 79 (2): 926–935.

4 West, A.M.A., Elber, R., and Shalloway, D. (2007). Extending molecular dynamics time scales with milestoning: example of complex kinetics in a solvated peptide. *J. Chem. Phys.* 126 (14): 145104.

Index

Molecular Kinetics in Condensed Phases: Theory, Simulation, and Analysis,
First Edition. Ron Elber, Dmitrii E. Makarov and Henri Orland.
© 2020 John Wiley & Sons Ltd. Published 2020 by John Wiley & Sons Ltd.